MYSTAT

STATISTICAL APPLICATIONS

DOS EDITION

Robert L. Hale
■ The Pennsylvania State University ■

Course Technology, Inc.
One Main Street
Cambridge, MA 02142

MYSTAT: Statistical Applications, DOS Edition is published by Course Technology, Inc.

Publisher	Joseph B. Dougherty
Project Manager	Katherine T. Pinard
Editorial Assistant	Nicole Jones
Production Manager	Josh Bernoff
Production Coordinator	Paddy Marcotte
Text Designer/Desktop Publisher	Debbie Crane Darrell Judd
Cover Designer	Darci Mehall
Manufacturing Coordinator	Mark Dec
Quality Assurance	Mark Valentine Rob Spadoni
Student Testers	Allyson Popson Stephanie Lataif

Trademarks

Course Technology and the open book logo are registered trademarks of Course Technology, Inc.

MYSTAT and SYSTAT are registered trademarks of SYSTAT, Inc.

Some of the product names used in this book have been used for identification purposes only and may be trademarks or registered trademarks of their respective manufacturers and sellers.

Disclaimer

Course Technology, Inc. reserves the right to revise this publication and make changes from time to time in its content without notice.

ISBN 1-878748-63-7 (includes software, 5¼-inch disks)
ISBN 1-878748-64-5 (includes software, 3½-inch disks)

Foreword

Dear MYSTAT User:

At SYSTAT, Inc. we understand that working with statistics isn't always easy–or even exciting. But we also know that the need for statistical analysis has become more prevalent than ever. Professionals in a wide range of fields such as psychology, sociology, biology, medicine, market research, environmental research, criminology, economics, and even manufacturing are relying on statistical analysis to better understand their data. And with better understanding comes better solutions and better management.

SYSTAT, Inc. has consistently maintained a special commitment to the academic community. We developed MYSTAT nearly five years ago as an instructional statistical software program to be used in introductory statistics courses. MYSTAT is currently used in hundreds of academic institutions by thousands of students worldwide.

There will always be a need for statistics. We are happy to get you started with *MYSTAT: Statistical Applications*, one in a series of statistical software products, produced by Course Technology, Inc. and SYSTAT, Inc. These texts are customized specifically for educational use.

We hope you enjoy this text and we look forward to working with you in the future. Enjoy your statistics!

Leland Wilkinson
President, SYSTAT, Inc.

Preface

When I first began learning statistics, the only substantial calculation tool available was the slide-rule. Because every slide-rule performed the same functions in the same manner and the simple mathematical operations could be taught in a few hours, statistics instructors consciously integrated this calculation instrument into class.

Next came the electronic calculator. Its operation was much easier than the slide-rule. Formerly tedious calculations were performed almost instantaneously. Calculators were swiftly adopted by statistics instructors and students and were integrated into both classroom lectures and textbooks.

The calculator and slide-rule only helped to solve the computation problems; they did little to help reveal the deeper meanings of statistical decision theory. Students had to understand the calculation formulas in order to be successful with the tool. Many students felt an overwhelming amount of time was devoted to the mathematics, and far too little time was spent on the assumptions and theoretical issues behind the statistical procedures. Often students could perform calculations on examinations, but they couldn't explain why one statistical procedure was used instead of another. Statistics class was often dreaded because of the math, and all excitement about the power that statistics provides to help make important research decisions for human welfare or business was lost.

With the proliferation of personal computers and statistical software programs, the learning scene has changed for the better. With current tools—a personal computer and a user-friendly statistics package—it is possible to answer statistical questions without spending inordinate amounts of time calculating the solution. To teach students how to do this successfully, however, it is necessary to integrate today's tools thoroughly into the classroom. With this integration, statistics instructors can concentrate on teaching the assumptions and theoretical issues of statistics, instead of teaching basic mathematical calculations. The tutorials and examples in this text enable the computer to become more than a simple computation tool; it can be a learning tool.

Content and Organization

MYSTAT: Statistical Applications is coordinated with the learning sequence presented in the typical introductory statistics text. With only a few exceptions, (for example, where there is a reference to data or an illustration presented in a previous chapter) the chapters are independent, and can be taught in any order.

Chapter 1 explains how to install MYSTAT, then briefly describes the MYSTAT commands.

Chapter 2 introduces the students to data entry. Students learn how to enter and save data, how to create new variables, how to find specific cases using an *If...then* expression, and how read a text file with MYSTAT.

Chapters 3 and **4** give an introduction to descriptive statistics using MYSTAT to graph single variables and to calculate descriptive statistics.

Chapter 5 shows how to use MYSTAT to calculate one-sample z- and t-tests. Although MYSTAT can not directly calculate one-sample tests, the chapter shows how to use the software to do these calculations. MYSTAT's ZIF function, used to find the probability given the z-value, and ZCF function, used to find the z-value given the probability, are also described. Although the paired t-test is used in one-sample situations, initial discussion of this procedure is not presented until chapter 6 (*Two-Sample Statistical Tests*), because the procedure is more involved than the single sample examples presented in this chapter.

Chapter 6 uses MYSTAT to calculate t-tests for both one- and two-sample situations.

Chapter 7 explains how to use MYSTAT to calculate ANOVA.

Chapter 8 gives an introduction to graphs. Proper graphic design is explained and students learn how to construct scatterplots.

Chapter 9 describes correlation and regression techniques.

Chapter 10 gives the students an introduction to ANCOVA and explains how to use MYSTAT's ANOVA function to calculate ANCOVA statistics.

Chapter 11 explains how to use MYSTAT to calculate non-parametric statistics.

Pedagogical Features

Six-Step Solution

The six-step solution appears in each chapter that discusses a statistical test. The six steps provide a consistent framework for students to use to find a solution to a given problem. The steps guide the student through constructing the null and alternative hypotheses, setting the probability, calculating the stastistic using MYSTAT, and interpreting and analyzing the results.

Decision Charts

Each chapter that introduces a statistical test includes decision charts. These charts are designed to help students decide which statistical test is required for the data and problem being studied.

Data Disk

In addition to the MYSTAT software, *MYSTAT: Statistical Applications* comes with a student data disk. All the data used in the text and in the exercises are provided on the disk. The students can spend their time studying the data instead of keying it in, and, because the data is already on disk, data entry errors are avoided.

End of Chapter Exercises

Exercises are provided at the end of each chapter. Some of the exercises use data that has been used in the chapter; other exercises use entirely new data. All are designed to give the students practice in using MYSTAT to describe and make decisions about data.

Acknowledgments

I would like to thank John Connolly, Kitty Pinard, and David Crocco from Course Technology for all their assistance. Without their vision of integrating technological advances into the college curriculum, this book would have been impossible. I would also like to thank Debbie Crane for her wonderful interior design and for implementing it so quickly. Special thanks go to Rob Spadoni, Mark Valentine, and Josh Bernoff for their help with the installation section. And finally, I would like to thank Allyson Popson and Stephanie Lataif for the many hours they spent working through chapter drafts making sure that the directions were clear, and the results both correct and understandable.

It is also a pleasure to articulate my appreciation to the following reviewers for their thoughtful comments: Tony Dubitsky, ASI, Market Research Inc., Lawrence Gall, Yale University, Rebecca German, University of Cincinnati, and C. Lincoln Johnson, University of Notre Dame.

Thanks also go to Leland Wilkinson, Mary Ann Hill, and Eve Goldman of SYSTAT, Inc. for their suggestions as the text was being developed.

Finally, I would like to thank my wife, my son, and my mother for all their support through the years. Without their encouragement and my father's many days of working overtime to sustain my undergraduate studies, I would not have been able to obtain any of my goals. Thanks Dad!

From the Publisher...

At Course Technology, Inc. we are very excited about bringing you, college professors and students, the most practical and affordable technology-related products available.

The Course Technology Development Process

Our development process is unparalleled in the higher education publishing industry. Every product we create goes through an exacting process of design, development, review, and testing.

Reviewers give us direction and insight that shape our manuscripts and bring them up to the latest standards. Every manuscript is quality tested. Students whose background matches the intended audience work through every keystroke, carefully checking for clarity, and pointing out errors in logic and sequence. Together with our own technical reviewers, these testers help us to ensure that everything which carries our name is error-free and easy to use.

Course Technology Products

We show both **how** and **why** technology is critical to solving problems in college and in whatever field you choose to teach or pursue. Our time-tested, step-by-step instructions provide unparalleled clarity. Examples and applications are chosen and crafted to motivate students.

The Course Technology Team

This book will suit your needs because it was delivered quickly, efficiently and affordably. In every aspect of our business, we rely on a commitment to quality and the use of technology. Every employee contributes. The names of all of our employees, each equity holders in the company, are listed below:

Stephen M. Bayle, Josh Bernoff, Jan Boni, Irene Brennan, Susan Collins, John M. Connolly, Rebecca Costello, Debbie Crane, David Crocco, Mark Dec, Yvette Delgado, Katie Donovan, Joseph B. Dougherty, Susan Feinberg, Lori Glass, Suzanne Goguen, David Haar, Deanne Hart, Nicole Jones, Matt Kenslea, Peter Lester, Laurie Michelangelo, Kim Munsell, Paul Murphy, Amy Oliver, Debbie Parlee, George J. Pilla, Katherine Pinard, Diana Simeon, Robert Spadoni, Kathy Sutherland, Mark Valentine

Brief Contents

Contents

MYSTAT Overview

MYSTAT is an interactive statistics and graphics package. With MYSTAT, users can compute most of the descriptive and inferential statistics typically covered in a two-semester college course. MYSTAT graphs aid in visualizing both one-dimensional and two-dimensional data. MYSTAT is available for IBM PC-compatible and Macintosh computers and VAX/VMS systems. This text describes MYSTAT on IBM PC-compatible computers (PCs). MYSTAT requires 320K of memory and either a hard disk drive or two floppy disk drives. It will run in both MS-DOS and PC-DOS environments. MYSTAT can handle up to fifty variables and 32,000 cases.

Objectives

At the end of this tutorial you should be able to

- Install MYSTAT
- Start MYSTAT
- Understand MYSTAT's major commands
- Obtain help with MYSTAT
- Quit MYSTAT
- Run the MYSTAT Demo program

■ Installing and Starting MYSTAT

This tutorial describes how to install MYSTAT. Using the MYSTAT Install Disk that comes with this text, you can install the MYSTAT software program and the data files that will be used in the tutorials. You must complete one of the installation procedures described below before you can use MYSTAT.

The MYSTAT Install Disk includes two compressed files, PROGDISK.EXE and DATADISK.EXE. PROGDISK.EXE expands into all the program files you need to run MYSTAT. DATADISK.EXE expands into all the data files you need to do the tutorials in this book. The Install Disk also includes batch files that install MYSTAT automatically on dual-drive or hard-disk systems. To install MYSTAT, follow the instructions that are appropriate for your system:

■ If you have a system with two disk drives and no hard disk, skip to "Dual-Drive Installation" below.

■ If you have a system with a hard disk and you want to install MYSTAT on the hard disk, skip to "Hard-Disk Installation" on page 3.

■ If you are working with a system that has MYSTAT already installed on a hard disk (for example, in a computer lab on campus) and only one floppy disk drive, skip to "Lab Installation with One Disk Drive" on page 4.

■ If you are working with a system that has MYSTAT already installed on a hard disk and two floppy disk drives, skip to "Lab Installation with Two Disk Drives" on page 5.

Dual-Drive Installation

To install and run MYSTAT on a dual-drive system with no hard disk, you will need the following:

■ A DOS System disk containing DOS and the files COMMAND.COM and CONFIG.SYS

■ The MYSTAT Install Disk that came with this package

■ Two blank, formatted floppy disks that are the same size as drive B.

☞   You must have a CONFIG.SYS file on your DOS System disk and it must contain a FILES setting. If the CONFIG.SYS file is missing, if the FILES setting is less than 20, or if the FILES setting is missing, you cannot run MYSTAT. See your instructor or consult the DOS manual that came with your computer.

To create MYSTAT program and data disks,

1. Start your PC by inserting your DOS System disk in drive A, closing the disk drive door, then turning on the computer. Enter the date and time if the computer requests them.

   You should now see the DOS prompt (A>).

2. Get two blank, formatted floppy disks. Label one *MYSTAT Program Disk*, and label the other one *MYSTAT Data Disk*.

☞ If you need help formatting a disk, see your instructor or consult your DOS manual.

3. Replace the disk in drive A with the MYSTAT Install Disk.

4. Type **A:** and press [**Enter**] to make A the default drive.

5. Type **TWODRIVE** and press [**Enter**] to start the installation process for a dual-drive system.

   Messages will appear on the screen telling you what to do. Follow each instruction as it appears on your screen:

6. Read the first screen of explanation. Press [**Enter**] to continue.

7. Insert the blank, formatted MYSTAT Program Disk in drive B and press [**Enter**].

   The batch file expands the MYSTAT program files and copies them onto your MYSTAT Program Disk. Messages appear on the screen as the files are copied and expanded. After the copying is completed, you see the message

   `MYSTAT PROGRAM DISK COPIED`

8. Replace the disk in drive B with the blank, formatted MYSTAT Data Disk and press [**Enter**].

   The batch file expands the MYSTAT data files and copies them onto your MYSTAT Data Disk. After the copying is completed, you see the message

   `MYSTAT DATA DISK COPIED`

9. Put the original MYSTAT Install Disk in a safe place.

10. If your disk drives are the same size, you are finished installing MYSTAT. If your disk drives are different sizes, you must copy your MYSTAT Program Disk to a disk the same size as drive A. For example, if drive A is a 5¼-inch drive and drive B is a 3½-inch drive, copy the MYSTAT Program Disk to a 5¼-inch disk.

Skip to the section "Starting MYSTAT" on page 6.

Hard-Disk Installation

If you have a hard-disk system with MYSTAT already installed on it, skip to the appropriate "Lab Installation" section on page 4 or 5. To install and run MYSTAT on a hard-disk system, you will need the following:

■ The MYSTAT Install Disk that came with this package

■ One blank, formatted floppy disk, labeled *MYSTAT Data Disk.*

☞ If you need help formatting a disk, see your instructor or consult the DOS manual that came with your computer.

To install MYSTAT on a hard disk,

1. Make sure the computer is on and DOS is running. You should see the DOS prompt (often C> or C:\>).

2. Insert the MYSTAT Install Disk in drive A.

3. Type **A:** and press **[Enter]** to make A the default drive.

4. Type **HARDDISK C:\MYSTAT** and press **[Enter]** to start the installation process for a hard-disk system. (If you want to install MYSTAT in a drive or directory other than C:\MYSTAT, substitute the correct drive letter and directory name in this command.)

   Messages will appear on the screen telling you what to do. Follow each instruction as it appears on your screen:

5. Read the first screen of explanation. Press **[Enter]** to continue.

   The batch file expands the MYSTAT program files and copies them onto the specified directory on your hard disk. Messages appear on screen as the files are copied and expanded. After the copying is completed, you see the message

   ```
   MYSTAT PROGRAM COPIED TO C:\MYSTAT
   ```

6. When you are prompted to do so, replace the disk in drive A with the blank, formatted MYSTAT Data Disk and press **[Enter]**.

   The batch file expands the MYSTAT data files and copies them onto your MYSTAT Data Disk. After the copying is completed, you see the message

   ```
   MYSTAT DATA DISK COPIED
   ```

7. Type **C:** and press **[Enter]** to make C the default drive.

8. Put the original MYSTAT Install Disk in a safe place.

Skip to the section "Starting MYSTAT" on page 6.

Lab Installation with One Disk Drive

If you're working in a lab where MYSTAT is already installed, you still need to create your own data disk. To use this installation option, your computer must have only *one* floppy disk drive. If your computer has two floppy disk drives, skip to "Lab Installation with Two Disk Drives" on page 5. To create the data disk, you need

■ A computer with MYSTAT already installed on the hard disk and only one floppy disk drive

■ The MYSTAT Install Disk that came with this package

■ One blank, formatted floppy disk, labeled *MYSTAT Data Disk.*

☞ If you need help formatting a disk, see your instructor or consult the DOS manual that came with your computer.

To create a MYSTAT Data Disk,

1. Make sure the computer is on and DOS is running. You should see the DOS prompt (usually C> or C:\>).
2. Insert the MYSTAT Install Disk in drive A.
3. Type **A:** and press [**Enter**] to make A the default drive.
4. Type **LABDISK** and press [**Enter**]. This runs a batch file that expands the MYSTAT data files and copies them onto your MYSTAT Data Disk.

   Messages will appear on the screen telling you what to do. Follow each instruction as it appears on your screen:
5. Read the first screen of explanation. Press [**Enter**] to continue.
6. When prompted to insert a disk for drive B, replace the original MYSTAT Install Disk with the blank, formatted MYSTAT Data Disk and press [**Enter**].

   WARNING: *Do not mix up the disks.*
7. When prompted to insert a disk for drive A, replace the MYSTAT Data Disk with the original MYSTAT Install Disk and press [**Enter**].

   Repeat steps 6 and 7 as many times as necessary until you are no longer prompted to switch disks.

   As the data files are expanded, messages appear on the screen. After the copying is completed, you see the message

   ```
   MYSTAT DATA DISK COPIED
   ```
8. Type **C:** and press [**Enter**] to make C the default drive.
9. Put the original MYSTAT Install Disk in a safe place.

Continue with the section "Starting MYSTAT" on page 6.

Lab Installation with Two Disk Drives

If you're working in a lab where MYSTAT is already installed, you still need to create your own data disk. To use this installation option, your computer must have two floppy disk drives. To create the data disk you need

- A computer with MYSTAT already installed on the hard disk and two floppy disk drives
- The MYSTAT Install Disk that came with this package
- One blank, formatted floppy disk, labeled *MYSTAT Data Disk.*

☞ If you need help formatting a disk, see your instructor or consult the DOS manual that came with your computer.

1. Copy the file DATADISK.EXE from the MYSTAT Install Disk to the blank, formatted MYSTAT Data Disk.

2. Be sure the MYSTAT Data Disk is in drive A.

3. Type **A:** and press **[Enter]** to make A the default drive.

4. Type **DATADISK** and press **[Enter]** to expand the MYSTAT data files on your MYSTAT Data Disk.

5. Type **C:** and press **[Enter]** to make C the default drive.

6. Put the original MYSTAT Install Disk in a safe place.

Skip to the section "Starting MYSTAT" below.

■ Starting Mystat

If you are starting MYSTAT from a dual-drive system, follow the instructions below. If you are starting MYSTAT from a hard-disk system, follow the instructions on the next page.

Starting MYSTAT on a Dual-Drive System

To start MYSTAT,

1. Start your PC by inserting your DOS System disk in drive A, closing the disk drive door, and turning on the computer. Enter the date and time if the computer requests them.

    You should now see the DOS prompt (A>).

2. Be sure the MYSTAT Program and Data Disks you created are not write-protected.

3. Replace the disk in drive A with the MYSTAT Program Disk.

4. Insert the MYSTAT Data Disk in drive B.

5. Type **B:** and press **[Enter]** to make B the default drive.

6. Type **MYSTAT** and press **[Enter]**. You see the copyright screen shown in Figure 1.1.

```
        ##########################################
        ###################################################
        ###    ###  ##     ##   ######  ########    ##   ########
        ####  ####  ##     ##  ###  ##     ##      ####      ##
        ## ## ## ##   ##  ##   ###         ##     ##  ##     ##
        ##    ##    ####     ###        ##    ########     ##
        ##    ##     ##    ##  ###    ##    ##    ##    ##
        ##    ##     ##   ######     ##   ##      ##  ##
        ###################################################
                 #############################################  TM
        Version 2.1              An Instructional Version of SYSTAT

 Copyright (c) 1991 Systat, Inc. All rights reserved worldwide. This software
 contains confidential and proprietary information of SYSTAT, Inc., which is
 protected by copyright, trade secret, and trademark law.  Copying or
 commercial distribution of this software is forbidden without permission from
 SYSTAT, Inc.

 This disk accompanies "MYSTAT: Statistical Applications," by Robert L. Hale,
 copyright (c) 1992 by Course Technology, Inc.  No part of the text may be
 reproduced in any form without written permission from the publisher.

 ┌─────────────────────────────┐
 │ Press ENTER ◄─┘or RETURN │
 └─────────────────────────────┘
```

Figure 1.1
MYSTAT copyright
screen

7. Press [**Enter**] to see the main command screen.

Continue with the section "MYSTAT Commands" on the next page.

Starting MYSTAT on a Hard-Disk System

To start MYSTAT,

1. Make sure the MYSTAT Data Disk you created is not write-protected and insert it in drive A.

2. Change to the drive and directory where the MYSTAT program is installed. For example, if MYSTAT is installed in C:\MYSTAT, type **C:** and press [**Enter**], then type **CD\MYSTAT** and press [**Enter**].

3. Type **MYSTAT** and press [**Enter**]. You see the copyright screen shown in Figure 1.1 above.

4. Press [**Enter**] to see the main command screen.

   ☞ The tutorials in this text are written for a PC with two floppy disk drives and no hard drive, and with the data disk in drive B. Substitute the letter of your floppy disk drive (usually A) for the B drive designation in the commands.

Continue with the section "MYSTAT Commands" on the next page.

■ MYSTAT Commands

After starting MYSTAT and pressing [Enter] to move past the copyright screen, you will see the main command screen shown in Figure 1.2.

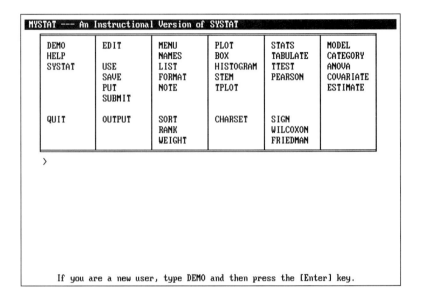

```
MYSTAT --- An Instructional Version of SYSTAT

    DEMO      EDIT      MENU      PLOT        STATS      MODEL
    HELP                NAMES     BOX         TABULATE   CATEGORY
    SYSTAT    USE       LIST      HISTOGRAM   TTEST      ANOVA
              SAVE      FORMAT    STEM        PEARSON    COVARIATE
              PUT       NOTE      TPLOT                  ESTIMATE
              SUBMIT

    QUIT      OUTPUT    SORT      CHARSET     SIGN
                        RANK                  WILCOXON
                        WEIGHT                FRIEDMAN

    >


       If you are a new user, type DEMO and then press the [Enter] key.
```

Figure 1.2
MYSTAT main
command screen

Statistical analyses are conducted from the main command screen. Data are entered in another screen called the MYSTAT Editor. (The MYSTAT Editor is a separate area in MYSTAT that will be discussed in Chapter 2.) The greater than sign (>) below the command list is the MYSTAT prompt. The cursor must be at the MYSTAT prompt when commands are keyed in.

The commands available in the main command screen are listed on the screen. HELP, QUIT, USE, SAVE, and FORMAT can also be used in the MYSTAT Editor. In addition, the following commands can be used only in the MYSTAT Editor: DELETE, DROP, FIND, GET, IF…THEN LET, LET, NEW, REPEAT.

The MYSTAT commands will be described in detail throughout the text. Table 1.1 is a quick reference to the commands and their functions.

ANOVA	used in conjunction with the CATEGORY and ESTIMATE commands to conduct analysis of variance
BOX	produces a box-and-whiskers plot for one or more variables
CATEGORY	identifies the independent variables in ANOVA and ANCOVA
CHARSET	an advanced command that allows users to choose either IBM or generic character sets for graphic output sent to the printer
COVARIATE	identifies the covariate in ANCOVA
DELETE	deletes an entire row of data (one case) in the Editor
DEMO	runs the MYSTAT Demo program
DROP	deletes an entire column of data (one variable) in the Editor

Table 1.1 cont.

EDIT	invokes the MYSTAT Editor where users create data files or edit previously created files
ESTIMATE	used in conjunction with the ANOVA and ANCOVA commands to conduct analysis of variance and analysis of covariance; used in conjunction with the MODEL command to conduct regression analyses
FIND	finds data values that meet specified criteria
FORMAT	determines the number of digits displayed to the right of the decimal point in numeric output or in the Editor; this number can vary between zero and nine
FRIEDMAN	conducts the Friedman two-way analysis of variance test
GET	reads a text file into MYSTAT
HELP	produces on-screen assistance with MYSTAT commands
HISTOGRAM	produces a histogram for one or more variables
IF...THEN LET	re-expresses the value of a variable if a specified condition is met
LET	re-expresses the value of a variable
LIST	lists the cases in the current data file
MENU	turns the MYSTAT menu on or off—MENU = ON turns the menu on (the default); MENU = OFF turns the menu off
MODEL	calculates regression lines; when used in conjunction with the ESTIMATE command conducts regression analyses
NAMES	prints the variable names of the data currently being used
NEW	clears the MYSTAT Edit window
NOTE	allows user notes to be placed in the output
OUTPUT	directs output to the screen (the default), the printer, or a disk file
PEARSON	conducts Pearson product moment correlations, or if the variables are ranked, conducts Spearman rank order correlations
PLOT	produces a bivariate scatterplot of one or more variables on the y-axis against a single variable on the x-axis
PUT	saves data to the disk as an ASCII file; .DAT is automatically appended to the end of each filename
QUIT	quits the MYSTAT Editor and returns the user to the main command screen from the Editor or terminates the MYSTAT program and returns the user to DOS from the main command screen
RANK	converts selected variables to their rank order
REPEAT	fills a specified number of cases in a new worksheet with missing values
SAVE	saves data created or edited with the data editor as MYSTAT files; .SYS is automatically appended to the end of each filename
SIGN	conducts the Sign test on pairs of variables
SORT	sorts data in ascending order using selected variables
STATS	produces descriptive statistics for selected variables; descriptive statistics can be produced for different groups within the data set if the data have been sorted using the grouping variable

Table 1.1 cont.

STEM	produces a stem-and-leaf diagram for numeric variables
SUBMIT	an advanced command that sends a command file to MYSTAT to process—command files are usually created by the user and contain lists of commands; files used with the SUBMIT command must end with a .CMD extension
SYSTAT	provides information about the differences between MYSTAT and SYSTAT, a commercial statistics package
TABULATE	produces one-way or multi-way tables, chi-square, and other associated statistics
TPLOT	plots a series of data values for one variable
TTEST	conducts dependent and independent *t*-tests
USE	specifies which data MYSTAT will use for the statistical analyses
WEIGHT	uses the integer value of a selected variable to weight cases
WILCOXON	conducts the Wilcoxon signed-rank test on pairs of variables

Table 1.1
MYSTAT commands

In this text, commands will be described in the format

COMMAND <option> [optional COMMAND]

The actual command is given in uppercase letters, but you may use uppercase and lowercase letters when typing the command. You may also abbreviate the command to its first two letters. The option given between less than and greater than signs (< >) is a variable. When you type the command, replace this word with the command, case, variable name, or variable on which you want to execute the command. Do not type the less than and greater than signs. Anything between brackets ([ ]) is an optional part of the command. If you use this part of the command, do not type the brackets.

■ Obtaining Help

The HELP Command

You can obtain on-screen help for MYSTAT commands by typing HELP followed by the command at the MYSTAT prompt. This displays information about the command you specify.

For example, do the following:

1. Type **HELP OUTPUT** and press **[Enter]**.

   You will receive the following help information:

```
OUTPUT    routes output to the screen, a file, or the printer.

OUTPUT    *           (sends subsequent output to screen only)
          @           (sends output to screen and printer)
        <file>        (sends output to screen and a file)

OUTPUT  FILE1       (sends output to FILE1.DAT)
```

2. Press [**Enter**] to return to the main command screen.

   Try another Help command.

3. Type **HELP STATS** and press [**Enter**].

   You will see the following information on your screen:

```
STATS prints descriptive statistics.  If you
choose no options, it produces N, MINIMUM, MAXIMUM, MEAN
and SD.  Otherwise, it prints just the option(s) you choose.
Use BY to get subgroup statistics after sorting the file by
the grouping variable(s). BY must follow any statistic options.

STATS [<var1>,<var2>,<...>]
   [/MEAN,SD,SKEWNESS,KURTOSIS,MINIMUM,MAXIMUM,RANGE,VARIANCE,SEM,SUM]
     [BY <var3>,<var4>,<...>]

STATS                     (basic statistics for whole file)
STATS VAR1,VAR2 / SEM     (standard error of the mean)
STATS / BY GROUP          (basic statistics for cases in each group)
```

   The first part of the output describes what the STATS command does. The next part of the output displays the entire STATS command with all the available options. Note that with the STATS command, you can select specific variables from the data, specify which descriptive statistics should be calculated, and specify if the results should be calculated and displayed by one or more grouping variables. Finally, in the last part of the help output, typical commands are given with explanations of results.

4. When you are finished looking at the HELP screen for the STAT command, press [**Enter**] to return to the main command screen.

   The HELP command is quite useful. You are encouraged to take a look at the help information for all the major commands at this point in the tutorial. To see the Help Index (a list of all the commands available from this screen) type **HELP** at the MYSTAT prompt and press [**Enter**].

Troubleshooting

If you type something that MYSTAT does not understand, MYSTAT gives you an error message. For example, if you are trying to have MYSTAT read a file but you misspell the filename, MYSTAT will give you the following message:

```
ERROR:
You are trying to read an empty file.
```

MYSTAT does not know that you misspelled the filename; it only knows that it cannot find the file you asked for.

The "Troubleshooting Guide" at the end of this book is an alphabetical guide to some of MYSTAT's error messages. (Error messages starting with the word ERROR are listed under the first letter of the next word.) A description of the error message along with a possible solution is given. If you encounter an error message, turn to the Troubleshooting Guide, look for the message you received, then follow the solution steps.

■ Quitting MYSTAT

When you are finished working with MYSTAT, quit the program and turn off the computer.

WARNING: *Students working in computer laboratories may be asked to leave the power on for the next student. Check with your lab instructor before continuing.*

1. Type **QUIT** at the MYSTAT prompt and press [**Enter**].

2. Shut the computer off.

3. Remove your floppy disks and take them with you when you leave.

■ The DEMO Command

MYSTAT contains a Demo program. This program demonstrates many of MYSTAT's features.

☞ The Demo program takes approximately five minutes and cannot be interrupted.

To run the Demo program,

1. With MYSTAT running, type **DEMO** at the MYSTAT prompt and press [**Enter**].

2. Press [**Enter**] after each screen is presented.

At the end of the demonstration, you will be returned to the main command screen. This command also creates a data set titled CITIES.SYS on the MYSTAT Program Disk or in the MYSTAT directory.

☞ In order to run the Demo program on a dual-drive system, you must start MYSTAT from drive A. At the DOS prompt, change the current drive to drive A by typing **A:** and pressing [**Enter**], then restart MYSTAT by typing **MYSTAT** and pressing [**Enter**]. After running the Demo program, you must restart MYSTAT from drive B following the instructions in the section "Starting MYSTAT on a Dual-Drive System" on page 6.

2

Data Editor

Statisticians study data, collections of objects. Quantities that change and can be measured are called variables. For example, the time spent by several individuals reading the first chapter in this tutorial is a variable. Variables may be *quantitative* (for example, age or weight) or *categorical* (for example, brands of cars or occupations). To conduct statistical calculations, variables need to be named so that their meaning and derivation can be easily remembered, and their values need to be recorded. Often, to discover statistical patterns new variables must be created mathematically or, more often, created from other variables already collected. Finally, variables need to be saved to conduct future analyses. MYSTAT enables users to do all of this using the MYSTAT Editor and the commands available within the MYSTAT Editor.

Objectives

At the end of this tutorial you should be able to

- Recognize the MYSTAT Editor
- Enter data into MYSTAT from the keyboard
- Edit data
- Save data to a disk as either a MYSTAT file or a text file
- Create new variables with the LET and IF...THEN LET commands
- Search for and find data values that meet specific conditions with the FIND command
- Read a text file
- Fill the MYSTAT Editor worksheet with a given number of null cases
- Understand the functions of the SORT, RANK, and WEIGHT commands
- Format data and analyses
- Send output to the screen, the printer, or a disk file

■ The MYSTAT Editor

The MYSTAT Editor screen is where data are entered into MYSTAT.

1. If you are using MYSTAT on a hard disk, follow the directions for starting MYSTAT on page 7. If you are using MYSTAT on a floppy disk, follow the directions for starting MYSTAT on page 6.

2. After viewing the copyright notice, press **[Enter]**.

   The main command screen that now appears (see Figure 2.1) lists all of the available command choices.

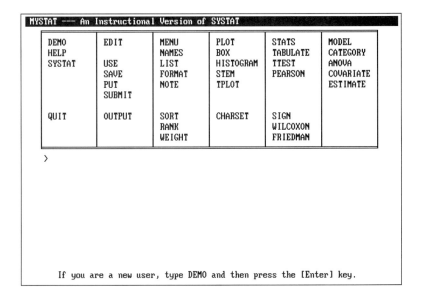

Figure 2.1
The main command screen

Below the command choices, you will see the MYSTAT *command line* and the MYSTAT prompt (>). All MYSTAT commands are typed on the command line after the MYSTAT prompt (>). The cursor is next to the MYSTAT prompt.

The EDIT command starts the MYSTAT Editor.

3. Type **EDIT** and press **[Enter]**.

   ☞ Remember, you can use uppercase or lowercase when entering commands.

   Immediately, you will be presented with a screen titled MYSTAT Editor (see Figure 2.2). This screen contains two sections. The top portion is the MYSTAT Edit window bounded on the top by the title bar and at the bottom by a wide bar below the number 15. Below this bar is the command area where you type in editing commands. Pressing the Escape [Esc] key toggles back and forth between these two areas.

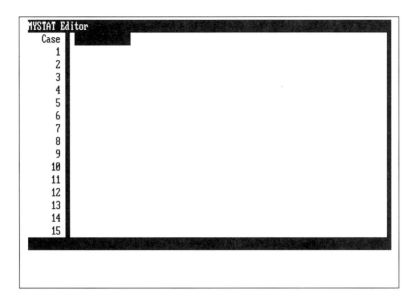

Figure 2.2
MYSTAT Editor

4. Press [**Esc**].

   The MYSTAT command prompt (>) and the cursor appear in the command area.

5. Press [**Esc**] a second time to return to the Edit window.

   The MYSTAT prompt and the cursor no longer appear in the command area, and an area in the Edit window is highlighted.

   To make a data set, you will store both variable names and variable values within the Edit window, which looks very much like a spreadsheet. The first column within the MYSTAT Edit window has numbers in every row with the exception of the first row, which is labeled *Case.* This first row, which we will call the *variable-naming row,* is where you are required to name each variable before being allowed to enter the variable values. Each subsequent row has a number in the first column. These numbers provide a convenient reference to the number of *cases*, or subjects, that are stored in the data set.

   Each intersection of a row and a column in the Edit window is called a *cell.* The black box indicates the current cell.

■ Keyboard Entry

The simplest method for entering data is from the keyboard. You will key in data that lists the 1985 populations (in thousands) and areas (in square miles) of the twenty largest cities in the world. You may be surprised to see how few U.S. cities are on the list. The data are listed in Table 2.1. Do not enter the data yet.

City	Population	Area
Tokyo	25434	1089
Mexico City	16901	522
Sao Paolo	14911	451
New York	14598	1274
Seoul	13665	342
Osaka	13562	495
Buenos Aires	10750	535
Calcutta	10462	209
Bombay	10137	95
Rio	10116	260
Moscow	9873	379
Los Angeles	9638	1110
London	9442	874
Paris	8633	432
Cairo	8595	104
Manila	8485	188
Jakarta	8122	76
Essen	7604	704
Teheran	7354	112
Delhi	6993	138

Table 2.1
Population and area
of the 20 largest
cities in the world

Naming Variables

There are three variables in this data set. The first is the name of the city. This type of variable is often called a *text*, *ASCII*, or *character* variable because the values are expressed with alphabetic characters. In MYSTAT, variable names always begin with a letter and can be up to eight characters long. Names of text variables must end with a $, which does not count as one of the eight characters.

The second variable, the population in 1985, must also be meaningfully abbreviated using eight or fewer characters. Since population is a *numeric* variable (the values are numbers) no special ending is required for the variable name. The third variable, area, is also numeric.

☞ When you enter text into the MYSTAT Editor, it must be preceded by a single (') or double quote (").

To name the variables for the population data,

1. Type **'CITY$** in the first column of the variable-naming row in the MYSTAT Editor, and press **[Enter]**.

   ☞ If you try to change a variable type after the variable name has been entered, MYSTAT gives you an error message. If you want to change the variable type after you have entered a variable name— for example, if you forgot the $ at the end of CITY—you must drop the entire column. See the section "Editing Data" on page 19.

2. Type **'POP** in the second column of the variable-naming row. Don't forget to press **[Enter]**.

3. Name the third variable by typing **'AREA** in the third column of the variable-naming row. Don't forget to press **[Enter]**.

   ☞ If you notice a typing error after you press **[Enter]**, use the cursor keys to go back to the cell containing the error and retype the entry.

Your MYSTAT Edit window should look like Figure 2.3.

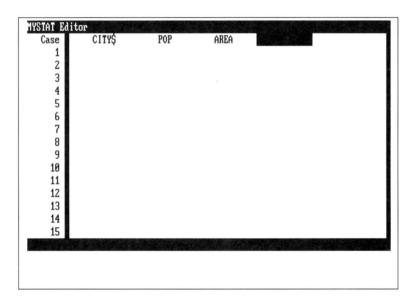

Figure 2.3
MYSTAT Editor with variables named

To enter values for the variables,

1. Press the **[Home]** key to place the cursor under CITY$ in the first numbered row. (If a 7 is typed when you press [Home], press **[Num Lock]**, delete the 7, and press **[Home]** again).

   The cell in the first column of the first numbered row will be highlighted. You may now type in the character value of that variable. Character values can be up to twelve letters long.

MYSTAT differentiates between upper and lowercase letters, so *Tokyo* and *tokyo* would be two distinct entries.

☞   If you don't have a value for a text variable, you may type in a double or single quote followed by a blank space and then press [**Enter**] to indicate the missing value.

2. Type **'Tokyo** and press [**Enter**].

MYSTAT enters this value and moves to and highlights the first cell under POP.

WARNING: *If your data disk is write-protected (see Chapter 1), you will get an error message. Remove the write-protect tab from a 5¼-inch disk or slide the write-protect tab down on a 3½-inch disk, then press* **R** *to Retry. You must work with a disk that is not write-protected.*

3. Type **25434** and press [**Enter**].

This value will be displayed as 25434.000. The MYSTAT Editor default display is three decimal places.

Numeric variables may not have more than twelve digits before or after the decimal place, and they may not have more than fifteen digits total. You must enter very large or small numbers using scientific notation. Variable values may not be larger than 10^{35}. If a numeric value is missing, you can type in a period (**.**) to indicate that it is missing.

The first cell under AREA should now be active.

4. Type **1089** and press [**Enter**].

You automatically are moved to the next active cell in the second row under CITY$.

5. Fill in the rest of the variable values using the values given in Table 2.1.

☞   Remember, if you encounter an error message, consult the Troubleshooting Guide.

When you enter the last number in the fifteenth row, the MYSTAT Editor will scroll so that you can enter the sixteenth case. You can move back to the top by using the [↑] (Up Arrow) key or the [Home] key. You can move right or left in the MYSTAT Editor by using the Right or Left Arrow keys. Table 2.2 lists the cursor movement keys and their functions.

☞   If your keyboard doesn't have cursor movement keys, you can use alternative key combinations. To see these alternative keys, press [**Esc**] to switch to the command line, then type **HELP** and press [**Enter**]. The cursor commands are listed on the top of the screen.

Key	Function
Home	Moves cursor to the first cell in the Editor
End	Moves the cursor to the last cell in the Editor
PgUp	Scrolls the Editor up one screen
PgDn	Scrolls the Editor down one screen
Ins	Scrolls the Editor one screen to the left
Del	Scrolls the Editor one screen to the right
Up Arrow	Moves the cursor up one cell
Down Arrow	Moves the cursor down one cell
Right Arrow	Moves the cursor one cell to the right
Left Arrow	Moves the cursor one cell to the left

Table 2.2
Cursor movement keys

6. Press **[Home]** to return to the top of the screen. Your completed data set should look like Figure 2.4.

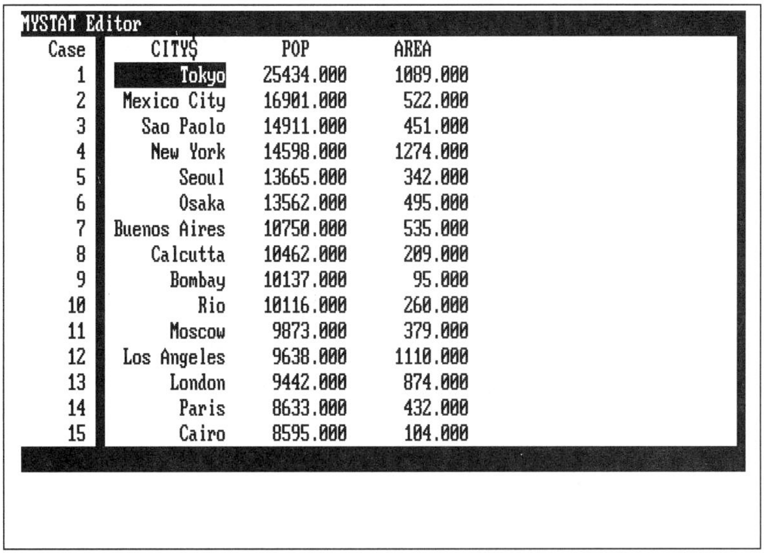

```
MYSTAT Editor
   Case      CITY$        POP       AREA
     1        Tokyo    25434.000  1089.000
     2   Mexico City   16901.000   522.000
     3    Sao Paolo    14911.000   451.000
     4     New York    14598.000  1274.000
     5        Seoul    13665.000   342.000
     6        Osaka    13562.000   495.000
     7  Buenos Aires   10750.000   535.000
     8     Calcutta    10462.000   209.000
     9       Bombay    10137.000    95.000
    10         Rio     10116.000   260.000
    11       Moscow     9873.000   379.000
    12  Los Angeles     9638.000  1110.000
    13       London     9442.000   874.000
    14        Paris     8633.000   432.000
    15        Cairo     8595.000   104.000
```

Figure 2.4
MYSTAT Editor with CITY data

■ Editing Data

If you notice an error in the data, move the cursor to the cell with the error, type the correct value, and press either [Enter] or the Up or Down Arrow key. The new value replaces the old value. If you make an error before you complete an entry, back up by pressing the Backspace key and enter the correct value.

You can delete an entire row by using the **DELETE <case>** command, and you can delete an entire column by using the **DROP <variable>** command. Be careful! Once the data is removed, it is gone forever.

To practice deleting a row, complete the following steps:

1. Press **[Esc]** to move to the command line in the MYSTAT Editor.

2. Type **DELETE 1** as shown in Figure 2.5, and press **[Enter]**.

```
MYSTAT Editor
   Case      CITY$        POP        AREA
     1        Tokyo    25434.000    1089.000
     2   Mexico City   16901.000     522.000
     3   Sao Paolo     14911.000     451.000
     4    New York     14598.000    1274.000
     5       Seoul     13665.000     342.000
     6       Osaka     13562.000     495.000
     7  Buenos Aires   10750.000     535.000
     8    Calcutta     10462.000     209.000
     9      Bombay     10137.000      95.000
    10        Rio      10116.000     260.000
    11      Moscow      9873.000     379.000
    12  Los Angeles     9638.000    1110.000
    13      London      9442.000     874.000
    14       Paris      8633.000     432.000
    15       Cairo      8595.000     104.000


>DELETE 1
```

Figure 2.5
MYSTAT Editor with
DELETE command

The first row where the information for Tokyo (Case 1) was stored is deleted.

When you delete a row, each row below the deleted row will move up one row. If you delete a column, each column to the right of the deleted column will move left one column. You cannot insert a row or column between existing rows or columns in the MYSTAT Editor, so you have to reenter the deleted data in the last row or column.

To reenter the Tokyo data,

1. Press **[Esc]** to move back into the Edit window from the command line.

2. Press **[End]** to move to the last case in the file.

3. Reenter the Tokyo data from Table 2.1 in the next row available (Case 20).

4. Press **[Home]** to move back to the first case in the file.

Deleting a column is just as easy as deleting a row. In the command line, type **DROP** followed by the variable name of the column you want to delete and press **[Enter]**. We won't do this now since you would have to reenter twenty values.

■ Saving Data

Data must be saved in a MYSTAT disk file before any statistical analyses can be conducted using the data. You should also always save your data after making changes. To save data, use the **SAVE <filename>** command. If you do not specify a path, the file is saved to the current drive and directory—the MYSTAT directory if you installed MYSTAT on a hard drive and the MYSTAT Program Disk in drive A if you are running MYSTAT from a floppy. It is recommended that you save all your data to the data disk. To save a file to a specific disk, specify the disk drive name followed by a colon and the filename. If you want to specify a path to store the file in a subdirectory, the entire path and filename must be enclosed in quotes. In this text, the assumption is that all files are contained or saved on the data disk in drive B, so the format with the drive letter B is used. If you have MYSTAT on a hard drive and the data disk in drive A, substitute the letter A for the letter B in the tutorial steps.

MYSTAT filenames must begin with a letter, may contain letters and numbers, and can be up to eight letters long. MYSTAT automatically gives data files the extension .SYS. Here are some examples of the SAVE command:

- **SAVE CITY** creates a MYSTAT data file called CITY.SYS on the current drive and directory.
- **SAVE '\DATA\FILE1'** saves the data in a MYSTAT data file named FILE1.SYS in the directory \DATA on the current drive.
- **SAVE B:CITY** saves the data in a MYSTAT data file titled CITY.SYS on the B drive. As noted above, this is the form used in this text.

To save the CITY data, use the following steps:

1. Make sure the data disk is in drive B.

2. Press **[Esc]** to move the cursor to the command line.

WARNING: *Remember, if you are working with MYSTAT on a hard disk, substitute the name of your floppy drive (usually A) for the letter B in the following step.*

3. Type **SAVE B:CITY** and press **[Enter]**.

   MYSTAT gives you a message telling you that twenty cases have been saved to a MYSTAT file (in this case, on the data disk).

If you try to save a new data set using the same name as a previous data set, you will be warned that the file you have named already exists and asked if you want to write over it. If you press **Y**, the older file will be erased and the new one will replace it.

☞ There may be occasions when you want to save your data in a text file instead of as a MYSTAT file; for example, you may want to import the data into a word-processing program. To save data as a text file use the **PUT <filename>** command while you are in the main command screen after saving the data from the Editor. (PUT does not work in the Editor.) To save the CITY data as a text file, you would type PUT B:CITY from the main command screen.

■ Transforming Data

It is often necessary to create new variables or transform (re-express) existing variables. Frequently a new variable must be generated from one or more previously stored variables. In this tutorial you will create two new variables by typing two new names in the variable-naming row. For the first new variable, you will use the **LET** command in the MYSTAT Editor to calculate all of the first variable's values. The LET command allows you to re-express the value of a variable (for example, square root the values). For the second variable, you will use the **IF...THEN LET** command in the MYSTAT Editor to create text values. The IF...THEN LET command allows you to add conditions *(if...then)* when you re-express values.

Using the LET Command

First, we'll calculate the number of people per square mile in each city. These values can be found by dividing each city's population (POP) by the city's area (AREA) and multiplying by 1,000.

To accomplish this, complete these steps:

1. Press [**Esc**] to return to the command line if you are not there. (Do not enter the LET command in the Edit window.)

2. Type **LET DENSITY = POP/AREA * 1000**, then press [**Enter**].

   The LET command automatically creates a new variable called DENSITY and fills that column with the calculated values. If the variable DENSITY had already existed, the LET command would have substituted the new, calculated values for the existing values. Your screen should look like Figure 2.6.

Notice that Bombay is the densest city shown in the window. If you scroll down using the arrow keys, you will see that Jakarta is even denser—almost ten times more people per square mile than New York!

```
MYSTAT Editor
  Case      CITY$       POP       AREA      DENSITY
     1   Mexico City   16901.000   522.000   32377.395
     2    Sao Paolo    14911.000   451.000   33062.084
     3    New York     14598.000  1274.000   11458.399
     4      Seoul      13665.000   342.000   39956.140
     5      Osaka      13562.000   495.000   27397.980
     6  Buenos Aires   10750.000   535.000   20093.458
     7    Calcutta     10462.000   209.000   50057.416
     8     Bombay      10137.000    95.000  106705.263
     9       Rio       10116.000   260.000   38907.692
    10     Moscow       9873.000   379.000   26050.132
    11  Los Angeles     9638.000  1110.000    8682.883
    12     London       9442.000   874.000   10803.204
    13      Paris       8633.000   432.000   19983.796
    14      Cairo       8595.000   104.000   82644.231
    15     Manila       8485.000   188.000   45132.979

>
```

Figure 2.6
CITY data after the
LET command is
executed

Using the IF…THEN LET Command

Suppose you are trying to decide which cities are "huge" and which are simply "big." You could generate another set of variable values to indicate your decision. The IF…THEN LET command allows the use of "*If* condition *then* action" clauses. The general form of this command is

IF <expression1> THEN LET <variable> = <expression2>

You are going to create a new character variable. After you have set up a proper *If* condition *then* action equation, MYSTAT will input the value *Big* if the population is less than 10 million. After constructing a second equation, MYSTAT will input *Huge* for the variable value if the population is equal to or greater than 10 million.

IF…THEN LET works with both character and numeric variables. For small data sets it might be faster to type in the new variable values yourself. If you have a large data set, it is faster and more efficient to let MYSTAT do the work. IF…THEN LET is typed on the command line, so before completing the following steps, press **[Esc]** to return to the command line if you are not already there.

You need to create an expression that says, "If POP is less than 10,000, then set SIZE$ equal to the word *Big*."

1. Type **IF POP < 10000 THEN LET SIZE$ = "Big"**.

   Note that character values must have quotation marks around them.

2. Press **[Enter]**.

You are now halfway through creating the values for the SIZE$ variable. Next you need to create the equation for the huge cities.

1. In the command line, type **IF POP >= 10000 THEN LET SIZE$ = "Huge"**.

2. Press [**Enter**].

The data should look like Figure 2.7.

```
MYSTAT Editor
   Case      CITY$        POP       AREA      DENSITY     SIZE$
      1   Mexico City  16901.000   522.000   32377.395     Huge
      2    Sao Paolo   14911.000   451.000   33062.084     Huge
      3    New York    14598.000  1274.000   11458.399     Huge
      4      Seoul     13665.000   342.000   39956.140     Huge
      5      Osaka     13562.000   495.000   27397.980     Huge
      6  Buenos Aires  10750.000   535.000   20093.458     Huge
      7    Calcutta    10462.000   209.000   50057.416     Huge
      8     Bombay     10137.000    95.000  106705.263     Huge
      9      Rio       10116.000   260.000   38907.692     Huge
     10     Moscow      9873.000   379.000   26050.132     Big
     11  Los Angeles    9638.000  1110.000    8682.883     Big
     12     London      9442.000   874.000   10803.204     Big
     13      Paris      8633.000   432.000   19983.796     Big
     14      Cairo      8595.000   104.000   82644.231     Big
     15     Manila      8485.000   188.000   45132.979     Big
```

Figure 2.7
The CITY data after completing the IF...THEN LET command

3. Type **SAVE B:CITY**, then press [**Enter**]. (You will be using this file later.)

MYSTAT gives you the following warning message in the command area:

```
***WARNING*** The file you have named already exists.
Do you want to write over it? (Y or N)
```

4. Type **Y**. (You do not need to press [Enter]).

■ Finding Specific Cases

Often in statistical research, cases with specific values for variables need to be found. Obviously, in a small data set like the one we just created, it is quite easy to look at the values to see if a particular value for a variable exists. However, for larger data sets, MYSTAT has a function that quickly finds values.

Suppose you wished to find all the cities with more than 100,000 people per square mile. Do the following:

1. Press **[Esc]** to go to the Edit window if you are not already there.

2. Go to the first cell in the MYSTAT Editor by pressing **[Home]**.

    This makes sure you start at the top of the data and will find all the cases.

3. Press **[Esc]** to return to the command line.

4. Type **FIND DENSITY > 100000**, then press **[Enter]**.

    Your screen should look like Figure 2.8.

```
MYSTAT Editor
 Case        CITY$         POP       AREA      DENSITY      SIZE$
    1    Mexico City    16901.000    522.000    32377.395     Huge
    2      Sao Paolo    14911.000    451.000    33062.084     Huge
    3       New York    14598.000   1274.000    11458.399     Huge
    4          Seoul    13665.000    342.000    39956.140     Huge
    5          Osaka    13562.000    495.000    27397.980     Huge
    6   Buenos Aires    10750.000    535.000    20093.458     Huge
    7       Calcutta    10462.000    209.000    50057.416     Huge
    8         Bombay    10137.000     95.000   106705.263     Huge
    9            Rio    10116.000    260.000    38907.692     Huge
   10         Moscow     9873.000    379.000    26050.132      Big
   11    Los Angeles     9638.000   1110.000     8682.883      Big
   12         London     9442.000    874.000    10803.204      Big
   13          Paris     8633.000    432.000    19983.796      Big
   14          Cairo     8595.000    104.000    82644.231      Big
   15         Manila     8485.000    188.000    45132.979      Big
```

Figure 2.8
CITY data after FIND command is executed

You asked MYSTAT to find the next case in which DENSITY is greater than 100,000 people per square mile. When you pressed [Enter], the first cell in the MYSTAT Editor that met this condition was highlighted. This city is Bombay. To find the next case, repeat steps 3 and 4.

■ Reading Text Files

Instead of typing data by hand, statisticians often read in data that have been saved by other researchers for other computer programs (often on other computer systems) in a text format. This format is frequently used to transfer data from one computer system to another. On the data disk, there is a relatively large data set saved in a text format. It is titled "IQTEXT.DAT."

Before you can read a text file, you must name the variables. There are thirteen numeric variables in the file IQTEXT.DAT. We'll give them temporary names until we find out more information about the data.

To name the variables,

1. With the cursor on the command line, type **NEW** and press **[Enter]**.

   The **NEW** command clears the Editor.

2. Press **[Esc]** to get into the Edit window.

3. Type **'A** to name the first variable, then press **[Enter]**.

4. Continue naming all thirteen variables by typing **'B, 'C, ... 'M.** When you're finished, your screen should look like Figure 2.9.

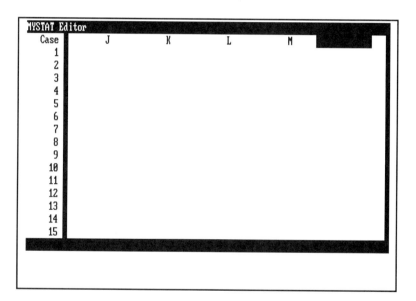

Figure 2.9
Temporary text file variable names

☞ If MYSTAT gives you the following error message:

    ERROR
    You have used up your EDITOR variables space.

you must quit the program and then restart it. See the Troubleshooting Guide for help.

5. After you have named the thirteenth variable (M), press **[Esc]** to get to the command line.

   Now you can read in the text file. To do this, you'll use the **GET <filename>** command.

6. Type **GET B:IQTEXT** to read in the text file IQTEXT stored on the data disk in drive B.

Depending on your computer system, it may take several minutes for MYSTAT to read in all the data. This data set contains psychological test information on 200 children referred for a psychological evaluation. By scrolling through this data set, you will discover that it has 200 cases.

When school children are referred for a psychological examination, they are given a battery of tests, including intelligence tests and achievement tests. Often behavior is also measured. Then the regular education teacher, special education teacher, parents, school administrator, and psychologist meet as a multidisciplinary team (MDT) to discuss the child. If the child needs special education, it is decided in this meeting. This data set contains information concerning 200 children after the MDT meeting was held.

The names of the variables you initially entered are not very descriptive, so you should give them meaningful names. The first variable is the group assignment the multidisciplinary team gave the child. Only numbers were recorded. Children assigned to group 1 were assessed as trainable mentally retarded. Group 2 children were reported as emotionally disturbed and withdrawn. Group 3 children are nonhandicapped. Group 4 children are learning disabled. Group 5 children are emotionally disturbed and aggressive. Group 6 children are bright average. There is only a single child in group 7.

Call the first variable MDT.

1. With the cursor in the first cell in the variable-naming row (A), type **'MDT**, then press [**Enter**].

   The second variable is the child's identification number. This variable is a number from 1 to 200.

2. The cell **B** should be highlighted. Type **'ID** to name this variable, and press [**Enter**].

   The third variable is the group assignment the school psychologist gave the child. The psychologist did not always agree with the MDT decision.

3. Type **'PSYCH** in cell **C**, then press [**Enter**] to name this variable.

   The fourth variable is the child's Verbal IQ. This variable measures how children respond to verbal questions.

4. In cell **D**, type **'VIQ** to name this variable.

5. Finish typing in each of the variable names as given below.

   The fifth variable is the child's Performance IQ (**PIQ**). Performance IQs measure how children perform on test questions that are more visual than verbal.

   The sixth variable is the child's Full Scale IQ (**FSIQ**). The FSIQ is a combination of the child's VIQ and PIQ.

   The next three variables are the child's reading, spelling, and arithmetic scores. You can abbreviate these variables **READ**, **SPELL**, and **ARITH**.

The tenth and eleventh variables are behavioral scores. The first variable indicates how many points the child received on an aggressiveness scale. Abbreviate this variable as **AGGRESS**. The second variable indicates how withdrawn the teacher perceives the child to be in the classroom. Abbreviate this variable as **WITHDRAW**.

The twelfth and thirteenth variables are the child's **AGE** (in months) and **GRADE** in school.

6. When you have finished editing this data set, press [**Home**] to return to the first column. Your file should look like Figure 2.10.

```
MYSTAT Editor
  Case      MDT        ID       PSYCH       VIQ        PIQ
   1       3.000     1.000      1.000      85.000     87.000
   2       2.000     2.000      2.000      98.000    100.000
   3       3.000     3.000      3.000      84.000     91.000
   4       1.000     4.000      4.000      65.000     64.000
   5       1.000     5.000      4.000      45.000     46.000
   6       3.000     6.000      1.000      92.000     96.000
   7       3.000     7.000      1.000      80.000     88.000
   8       3.000     8.000      4.000      64.000     75.000
   9       3.000     9.000      5.000     103.000    105.000
  10       3.000    10.000      5.000      92.000     88.000
  11       6.000    11.000      6.000     115.000    108.000
  12       3.000    12.000      1.000      82.000     93.000
  13       3.000    13.000      1.000      86.000     77.000
  14       1.000    14.000      4.000      60.000     71.000
  15       3.000    15.000      3.000      55.000     72.000
```

Figure 2.10
Completed text file in the MYSTAT Editor

Finally save the data set to the data disk in MYSTAT format. You will be using this data set in later chapters.

7. Press [**Esc**] to activate the command line.

8. Type **SAVE B:SCHOOL** and press [**Enter**] to save the data to the disk in MYSTAT format.

■ Odds and Ends

Functions, Relations, and Operators

There are many mathematical functions, relations, and operators you can choose to use within the commands LET, IF...THEN LET, and FIND. We did not discuss them all, but you should know what they are. Below is a quick definition for each.

+	addition
-	subtraction
*	multiplication
/	division
^	exponentiation
>	greater than
<	less than
=	equal to
<>	not equal to
>=	greater than or equal to
<=	less than or equal to
ABS	absolute value
ACS	arccosine
AND	logical and
ASN	arcsine
ATH	hyperbolic arctangent (Fisher's Z)
ATN	arctangent
CASE	current case number
COS	cosine
EXP	exponential function
INT	integer truncation
LOG	natural logarithm (base e)
OR	logical or
SIN	sine (argument in radians)
SQR	square root
TAN	tangent
URN	uniform random number (0,1)
ZCF	standard normal cumulative density function
ZIF	inverse normal cumulative density function
ZRN	normal random number (0,1)

Filling the Worksheet

Sometimes you may need to calculate a statistic and you don't have a data set to enter into MYSTAT. The **REPEAT** <#> command in the MYSTAT Editor allows users to fill a specified number of cases in a new worksheet with missing data values; later these missing values can be replaced with other values using the LET or IF...THEN LET commands.

To do this,

1. At the MYSTAT prompt on the command line, type **NEW** to clear the MYSTAT Editor.
2. Press [**Esc**] to move to the Edit window.
3. Type **'SCORE** and press [**Enter**] to name the first variable.
4. Press [**Esc**] to move back to the command line.
5. Type **REPEAT 25** and press [**Enter**].

   Your screen should look like Figure 2.11.

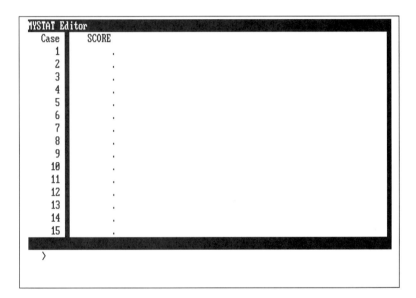

Figure 2.11
SCORE filled with
missing data values

The MYSTAT Editor filled the SCORE variable with missing data (indicated by decimal points) for twenty-five cases. You could now use the LET or IF…THEN LET command to have these missing data values changed to numbers.

The SORT, RANK, and WEIGHT Commands

The **SORT <variable>** command sorts cases by one or more variables and creates a new sorted file. The cases are sorted in either alphabetical or ascending order (smallest numbers first). MYSTAT also allows for nested sorts. This means you can sort a list by one variable, then subsort it by another, then by another, and so on. You may select up to ten numeric or text variables on which to do nested sorts. The sorted data are automatically saved in a new file you name on the disk. (To edit or perform statistical analyses on the sorted file, you will need to open it in MYSTAT.)

To sort a file, it is not necessary to load the file into the MYSTAT Editor. The **USE <filename>** command opens the file in MYSTAT to allow you to perform statistical analyses on it, but it does not load the file into the MYSTAT Editor.

Let's sort the CITY data to find the least and most densely populated cities.

To do this,

1. Type **QUIT** and press **[Enter]** at the MYSTAT prompt to get to the main command screen. You may get a message instructing you to save your work. If you have data to save, follow the instructions MYSTAT gives you; otherwise, press **[Enter]** again.

2. Type **USE B:CITY** and press **[Enter]** to load the CITY file from the disk in the B drive.

   A screen appears showing the names of the variables in this file.

3. Press **[Enter]** to return to the main command screen.

   Now you will create a file in which the sorted data will be stored.

4. Type **SAVE B:CITYSORT** and press **[Enter]**.

5. Type **SORT DENSITY** and press **[Enter]** to sort the file based on the DENSITY variable. Press **[Enter]** again when MYSTAT prompts you.

   Now look at the sorted file.

6. Type **EDIT B:CITYSORT** and press **[Enter]**.

   Your screen should look like Figure 2.12.

```
MYSTAT Editor
 Case      CITY$         POP        AREA      DENSITY      SIZE$
    1   Los Angeles    9638.000    1110.000    8682.883      Big
    2        Essen     7604.000     704.000   10801.136      Big
    3       London     9442.000     874.000   10803.204      Big
    4      New York   14598.000    1274.000   11458.399     Huge
    5        Paris     8633.000     432.000   19983.796      Big
    6  Buenos Aires   10750.000     535.000   20093.458     Huge
    7        Tokyo    25434.000    1089.000   23355.372     Huge
    8       Moscow     9873.000     379.000   26050.132      Big
    9        Osaka    13562.000     495.000   27397.980     Huge
   10   Mexico City   16901.000     522.000   32377.395     Huge
   11     Sao Paolo   14911.000     451.000   33062.084     Huge
   12          Rio    10116.000     260.000   38907.692     Huge
   13        Seoul    13665.000     342.000   39956.140     Huge
   14       Manila     8485.000     188.000   45132.979      Big
   15      Calcutta   10462.000     209.000   50057.416     Huge
```

Figure 2.12
CITY data sorted by DENSITY

7. Use the **[Home]**, **[End]**, **[PgUp]**, **[PgDn]**, and the arrow keys to scroll through the file.

You can see that Los Angeles is the least densely populated city and Jakarta is the most densely populated.

The **RANK <variable>** command produces a new data file in which the values of the chosen variables are replaced by their rank order. Tied ranks for selected variables are averaged. As with the SORT command, to use the RANK command, you must first open the file in MYSTAT with the USE command, then create a file in which the ranked data will be stored with the SAVE command. Also, as with the SORT command, you can rank the data based on one or more variables. You will use the RANK command in Chapter 9.

The **WEIGHT <variable>** command replicates cases by using the integer portions of the values of a weighting variable you select. The weighted data are used until you type WEIGHT again without any weighting variable. This command does not change the data file, but it does affect subsequent statistical analyses. For example, if you entered the data set in Table 2.3 and chose the variable NUMBER as the weighting variable (WEIGHT NUMBER), the SCORE of 1 would be replicated nine times and the SCORE of 2 would be replicated eight times in any analysis.

SCORE	NUMBER
1	9
2	8
3	7
4	6

Table 2.3
Sample data

Formatting Input

MYSTAT lets you change the number of decimal places displayed in the MYSTAT Editor. The general form of the command is **FORMAT <#>**. The number of decimal places displayed can be 0 through 9. The default is 3.

For example, if you are in the MYSTAT Editor with the CITYSORT file loaded,

1. At the MYSTAT prompt on the command line, type **FORMAT = 5**, then press [**Enter**].

   Your screen should look like Figure 2.13. The number of decimal places in the numeric variables has been changed from three to five.

Each cell can display only ten digits, so if you have the value 209.00 and give the command **FORMAT = 8**, this number will automatically be displayed in scientific notation.

This FORMAT command affects only what you see on the screen, not the precision of the data stored in the computer. MYSTAT stores at least fifteen digits for each number.

MYSTAT Editor

Case	CITY$	POP	AREA	DENSITY	SIZE$
1	Los Angeles	9638.00000	1110.00000	8682.88288	Big
2	Essen	7604.00000	704.00000	10801.13636	Big
3	London	9442.00000	874.00000	10803.20366	Big
4	New York	14598.00000	1274.00000	11458.39874	Huge
5	Paris	8633.00000	432.00000	19983.79630	Big
6	Buenos Aires	10750.00000	535.00000	20093.45794	Huge
7	Tokyo	25434.00000	1089.00000	23355.37190	Huge
8	Moscow	9873.00000	379.00000	26050.13193	Big
9	Osaka	13562.00000	495.00000	27397.97980	Huge
10	Mexico City	16901.00000	522.00000	32377.39464	Huge
11	Sao Paolo	14911.00000	451.00000	33062.08426	Huge
12	Rio	10116.00000	260.00000	38907.69231	Huge
13	Seoul	13665.00000	342.00000	39956.14035	Huge
14	Manila	8485.00000	188.00000	45132.97872	Big
15	Calcutta	10462.00000	209.00000	50057.41627	Huge

>

Figure 2.13
Formatted CITYSORT file

Formatting Output

MYSTAT allows you to format numeric output. To format numeric output, use the FORMAT command in the menu screen before any analyses are done. The form of the command is identical to the command in the MYSTAT Editor.

Directing Your Results

The OUTPUT command routes output to the screen, a file, or the printer. This command has three forms. **OUTPUT *** sends subsequent output to the screen. **OUTPUT @** sends subsequent output to the screen and to the printer. **OUTPUT <filename>** sends output to the screen and to a disk file. For example, **OUTPUT FILE1** would store subsequent output to a disk file titled FILE1.DAT. This command is important if you are working with a word processor. Since all word processors can read text files, it will be quite easy to read this file and copy and paste it into the results section of your working paper.

The default option is to send all output to the screen. To print graphs and statistical output, OUTPUT @ must be chosen before analyses and graphs are done. If you do an analysis first and then choose OUTPUT @, nothing will happen.

Exercises

1. Figure 2.14 is a small data set containing information about the planets. Enter this data by keying it in. (Remember to type **NEW** at the MYSTAT prompt to clear the Edit window.) The variables indicate the name of the planet, its DISTANCE from the sun, how many days it takes the planet to REVOLVE around the sun, and the DIAMETER of the planet.

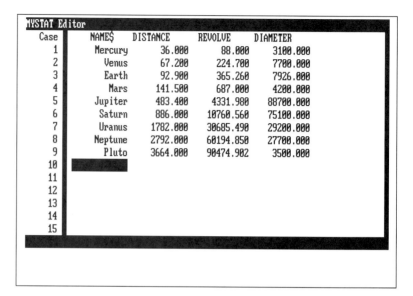

Figure 2.14
PLANETS data

2. Create a new variable that measures the time it takes each planet to revolve around the sun divided by the distance each is away from the sun. Is this value a constant? Save this data set as PLANETS.

3. Rank order the PLANETS data using both DISTANCE and DIAMETER as ranking variables. Save the ranked file as PLNTRANK. Open the PLNTRANK file and describe how the RANK command changed the data set.

4. Open the PLANETS file. Sort the data using NAME$ as the sorting variable. Save the file as PLNTSORT. Open the PLNTSORT file and describe how the SORT command changed the data set.

5. Data tabulated by Richard Heede of the Rocky Mountain Institute concerning federal spending on energy was reported in the March/April 1991 issue of *Sierra* magazine and has been saved on the data disk in a file titled ENERGY. This data set lists energy in the United States from various sources and the money the federal government gave in subsidies to each source.

The variables are defined as follows: SOURCE$ is the energy source. BTU is the number of quadrillion British Thermal Units (a unit of heat) produced or saved by each source. SUBSIDY is the dollars in billions the federal government gives in subsidies to each source. Finally, BTU_DOL is the number of BTUs each source provides per million dollars of expenditure by the government. Type in a new variable called DECISIN$ and use the following two statements to code the variable:

If BTU_DOL is less than 1.0, then set DECISIN$ to the word *Cut.*

If BTU_DOL is greater than or equal to 1.0, then set DECISIN$ to the word *Keep.*

You have just made decisions concerning energy sources to keep or cut from the federal budget. Be sure to save these data.

3

Graphing Data
(Single Variables)

If a picture is worth a thousand words, a graph may be worth a thousand calculations when trying to summarize a large data set. Often, the human eye can detect subtleties within a graph that no amount of statistical analysis can disclose. The challenge for the researcher is to produce accurate pictorial representations of data. If done correctly, graphs summarize and greatly speed our understanding of data. If done incorrectly, graphs can tell quite fanciful tales. MYSTAT provides several types of graphs that are helpful for visually representing single variables.

Objectives

At the end of this tutorial you should be able to

■ Produce and print histograms, stem-and-leaf diagrams, and box plots
■ Interpret typical scores and general shapes (skewness) of graphs
■ Produce and print TPLOT or series graphs
■ Understand the basics of good graphic design

■ What Are Graphs?

A graph is a pictorial representation of one or more variables. Graphs are used to view and understand the shape of the *distribution* (frequency of all the values) of a variable. They also are used to visualize the relation between two or more variables. This chapter focuses on one-variable graphs. Chapter 8 discusses two-variable graphs.

■ Histograms

A *histogram* is a graphical representation of the *density* or frequency of a single quantitative variable. It displays along the *x*-axis either single values or groups of values (*class intervals*). The height of each section of the histogram represents the frequency with which that value or class interval occurs. Unlike *bar charts*, in which the *x*-axis displays categories, the horizontal *x*-axis of a histogram shows a numeric scale.

A histogram is typically used to display the shape of a variable. However, a histogram's appearance can change depending upon the number of bars used. MYSTAT automatically chooses the best number of bars for revealing a variable's distribution. You may change this number if you wish.

Producing and Printing Histograms

Although a printout is often important, many reams of paper are wasted by indiscriminate printing. Wholesale printing is not necessarily helpful to understanding the data analysis, may be counterproductive (it takes time to print), and is objectionable on ecological grounds. Print only what you are assigned and only when necessary. Turn the print command off as soon as possible.

The HISTOGRAM command produces a histogram for each numeric variable. The general form of the HISTOGRAM command is:

HISTOGRAM [<VAR1>, <VAR2>, <...> / BARS = <#>, SCALE, MIN = <#>, MAX = <#>]

The BARS option is used to indicate how many bars you want in the histogram. The number of bars chosen by MYSTAT is exemplary and users are unlikely to do better. The SCALE option sets the *x*-axis to the nearest round numbers outside the data minimum and maximum. The MIN and MAX options allow you to set the minimum and maximum values along the *x*-axis. You do not need to use these options; MYSTAT selects a scale for the *x*-axis for you. The following commands are all valid forms of the HISTOGRAM command:

- **HISTOGRAM**   produces a histogram of every numeric variable in the data file.
- **HISTOGRAM X**   produces a histogram of the variable named X.
- **HISTOGRAM TRIAL(1), TRIAL(2)**   produces a histogram for the variable named TRIAL(1) and a second histogram for the variable TRIAL(2).

- **HISTOGRAM TRIAL(1-5)** produces a histogram for each of the variables TRIAL(1), TRIAL(2), TRIAL(3), TRIAL(4), and TRIAL(5).
- **HISTOGRAM A, B/BARS = 18** produces histograms for the variables A and B with each histogram having eighteen bars.
- **HISTOGRAM SCALE** produces histograms of every numeric variable in a file with the scale of the *x*-axis set so that the minimum and maximum values are the nearest round numbers outside the data minimum and maximum.
- **HISTOGRAM/MIN = 0, MAX = 20** produces a histogram with the scale of the *x*-axis set so that the minimum value is 0 and the maximum value is 20.

Let's produce and print the distribution of the population and density variables in the CITY data.

1. Start MYSTAT and press [**Enter**] to go from the copyright screen to the main command screen.

2. With the data disk in drive B, type **USE B:CITY**, then press [**Enter**].

   ☞   Remember, if you are running MYSTAT from a hard disk, your data disk is in drive A, and you should type **USE A:CITY**.

This loads the CITY file from the data disk in drive B into MYSTAT so that analyses can be conducted. The variable names from the CITY data appear on the screen as shown in Figure 3.1.

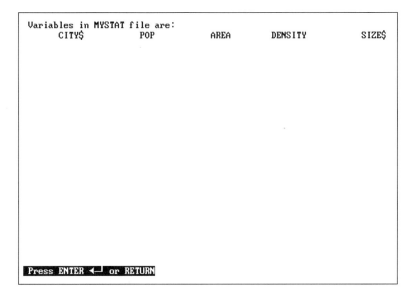

Figure 3.1
The variable name screen

3. Press [**Enter**] to return to the main command screen.

4. Type **OUTPUT @**, then press **[Enter]** to send the histograms to the screen and a printer attached to your computer.

WARNING: *If you do not want to print, omit step 4. See the section, "Directing Your Results" on page 33 in Chapter 2 for more information.*

5. Type **HISTOGRAM POP, DENSITY**, and press **[Enter]**. After the first histogram displays, press **[Enter]** to see the next one.

Your histograms should look like those shown in Figure 3.2.

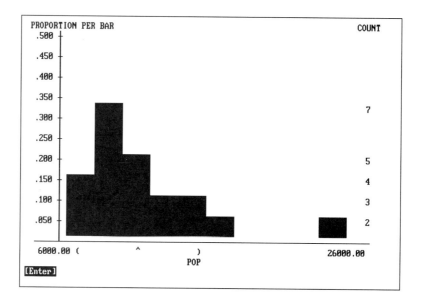

Figure 3.2a
Histogram of POP

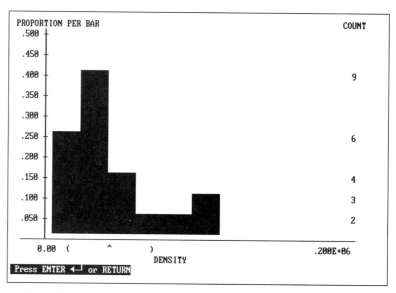

Figure 3.2b
Histogram of DENSITY

6. Press **[Enter]** to return to the main command screen.

Among other things, histograms show the *skew* of a distribution, which is a measure of its symmetry. If a distribution is symmetric, the left side of the distribution is a mirror image of the right side. If the values in a distribution bunch up at the smaller values, the distribution is positively skewed. If the distribution's values are predominantly at the larger end of the scale, the distribution is negatively skewed. See Figure 3.3.

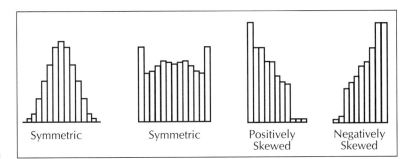

Figure 3.3
Skew of a distribution

In the histogram of populations (Figure 3.2a), the values of POP are scaled on the *x*-axis. MYSTAT labels the *x*-axis with the minimum and maximum values of the data (6000.00 and 26000.00), and indicates the location of the mean with an up carat (^) and the locations of the standard deviations with left and right parentheses. The numbers on the left *y*-axis show the proportion of scores in each interval. Multiplying this proportion by 100 yields a percentage. In Figure 3.2a, the proportion-per-bar for the first interval is over .150 or 15%. The *y*-axis on the right shows the number of cases per bar. The tallest bar contains seven cities. Since there are 20 cities, this is 35% of the sample. In the DENSITY histogram, the rightmost bar indicates cities with a large density. The right *y*-axis shows that three cities fall into this category. Also note that both histograms are positively skewed.

■ Stem-and-Leaf Graphs

Stem-and-leaf displays are like histograms turned on their sides except that all or part of the values for each variable appear in the graph. To show how this is done, assume you have the following values for a variable:

[10, 11, 12, 13, 15, 17, 20, 22, 23, 25, 27, 30, 33, 33, 33, 34, 36, 38, 39]

These numbers can be broken into two parts called the *stem* and the *leaf*. For these values, the stem will be the first digit and the leaf the second digit. Thus, the score of 10 has a 1 as its stem and a 0 as its leaf. For the score 27, the 2 would be the stem and the 7 would be the leaf.

If you produced a histogram of the data, it would look like Figure 3.4.

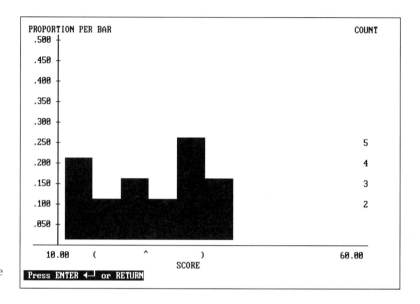

Figure 3.4
Histogram of variable
SCORE

Each class interval is a bar in a histogram, and each of these bars becomes a row in a stem-and-leaf display. Each row in a stem-and-leaf display is headed by its stem, and its length is determined by the number of leaves on the stem. For example, in creating the first row, we want to include all the numbers from 10 through 14 (10, 11, 12, 13). The stem for each of these numbers is 1 followed by the leaves 0123. The length of this row, which corresponds to the height of the bar in the histogram, is determined by the number of leaves following the stem. If the numbers from 15 through 19 are used for the second row, the stem will again be 1 and the leaves, 5 and 7. The only numbers within that class interval are 15 and 17. Note that each initial digit is used as a stem twice. Thus, the first class interval represented by the first bar in the histogram extends from 10 through 14, the second from 15 through 19, the third from 20 through 24, and so on. Taking this to its conclusion, you get a stem-and-leaf display like that in Figure 3.5. Histogram bars have been superimposed over the leaves to show the similarity to a histogram. Note that if you rotate the stem-and-leaf diagram 90 degrees to the left, you see that the leaves look just like the histogram in Figure 3.4.

The advantage of the stem-and-leaf display over a histogram is that additional information (the actual variable values) is contained in the display. Additionally, MYSTAT will calculate the first, second, and third quartile scores, and mark the rows within which each occurs with a special symbol. The first and third quartile scores (the 25th and 75th percentile ranks) are also known as *hinges* and so an *H* is placed within the space between the stems and leaves to indicate the row in which these quartile scores fall. The second quartile is also known as the *median* (the score at the 50th percentile) and is marked with an *M*. If you enter the data above and produce the full stem-and-leaf output from MYSTAT, you would obtain the output shown in Figure 3.6.

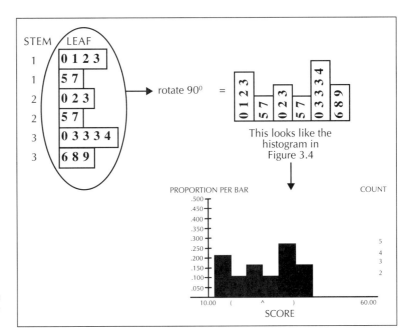

Figure 3.5
Relationship between
stem-and-leaf display
and a histogram

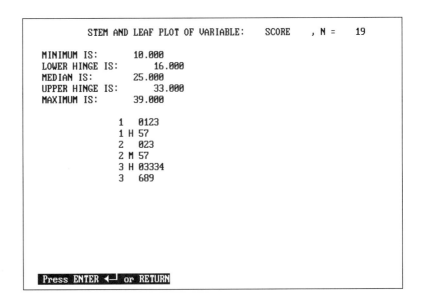

Figure 3.6
MYSTAT stem-and-
leaf display

The STEM command is used in the main command screen to produce stem-and-leaf plots. The general form for the STEM command is

STEM [<VAR1>, <VAR2>, <...>/LINES = <#>]

Note that it is similar to the HISTOGRAM command, except that the LINES option (instead of the BARS option) is used to indicate how many stems you want in the stem-and-leaf plot. Again, MYSTAT does a good job of making that decision, so this option is rarely used. Also, note that you cannot set a scale for a stem-and-leaf plot.

Using the CITY data, stem-and-leaf plots can be produced for the population and density variable.

1. Type **USE B:CITY**, and press [**Enter**] from the main command screen. Press [**Enter**] a second time to return to the main command screen from the variable name screen.

   ☞ If CITY is already loaded, you may skip this step.

2. Type **OUTPUT @**, and press [**Enter**] to direct output to the screen and the printer. (If you don't want to print, omit this step.)

3. Type **STEM POP, DENSITY** and press [**Enter**] to display the two stem-and-leaf plots.

   ☞ Remember to press [**Enter**] after viewing the first screen, then press [**Enter**] again to return to the main command screen.

The two stem-and-leaf plots shown in Figure 3.7 will be generated. Note that variables that are distant from the median are marked as outside values (see the stem-and-leaf displays). This is a useful measure for identifying values that stand apart from others. This frequently happens on one end when the distribution of a variable is seriously skewed. We will discuss how to determine when a data value is outside the expected values in the next section on box plots.

```
              STEM AND LEAF PLOT OF VARIABLE:        POP    , N =    20

MINIMUM IS:        6993.000
LOWER HINGE IS:       8540.000
MEDIAN IS:         9994.500
UPPER HINGE IS:      13613.500
MAXIMUM IS:       25434.000

                    6    9
                    7    36
                    8 H  1456
                    9 M  468
                   10    1147
                   11
                   12
                   13 H  56
                   14    59
                   15
                   16    9
               ***OUTSIDE VALUES***
                   25    4
```

Figure 3.7a
Stem-and-leaf of POP

```
                STEM AND LEAF PLOT OF VARIABLE:  DENSITY    , N =   20

   MINIMUM IS:      8682.883
   LOWER HINGE IS:    20038.627
   MEDIAN IS:      32719.739
   UPPER HINGE IS:    50365.665
   MAXIMUM IS:    106868.421

            0    8
            1    0019
            2 H  0367
            3 M  2389
            4    5
            5 H  00
            6    5
            7
            8    2
         ***OUTSIDE VALUES***
            10   66
```

Figure 3.7b
Stem-and-leaf of
DENSITY

4. Type **OUTPUT** * and press [**Enter**] to redirect output back to
the screen only.

■ Box Plots

Box or box-and-whiskers plots display the distribution of a single variable.
Figure 3.8 shows a typical box plot.

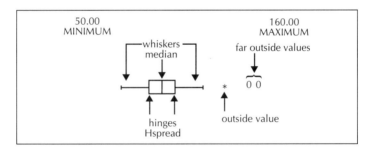

Figure 3.8
Typical box plot

In a box plot, the median of a variable is marked by a single vertical line. In
this diagram, the median has a value of approximately eighty-three. The lower
and upper hinges are the first and third quartiles (25th and 75th percentile
ranks) respectively, and the distance between them is called the *Hspread*. The
vertical lines that mark the hinges and the median are connected by horizontal
lines to form a box. The length of the box (the Hspread) is equivalent to the
interquartile range, that is, fifty percent of the values fall within the box. The
whiskers show how far the data spreads away from the hinges to a maximum
distance of 1.5 Hspreads. If data values do not spread all the way to ±1.5
Hspreads from the hinges, the whiskers do not extend that far. Whiskers
represent actual data points. Data values outside of the whiskers but less than
three Hspreads from the hinges are marked with an asterisk. Data values more
than three Hspreads from the hinges are marked by zeros.

The **BOX** command produces box-and-whiskers plots. The general form is
BOX [<VAR1>, <VAR2>, <...> * <VAR3>/GROUPS = <#>, MIN = <#>, MAX = <#>]

The * <VAR3> portion of the command produces *grouped* box plots. This option is used if the variable reflects subgroups of the full data; for example, if your data had scores for both males and females and a SEX variable coded these groups, you could produce separate box plots for males and females by substituting SEX for this option. The GROUPS option is used to indicate how many groups you want to be plotted in a group box plot. Exercise 2 at the end of this chapter asks you to make grouped box plots. The MIN and MAX options are used the same as in the HISTOGRAM command. They can be used to set the minimum and maximum values of the box plot scale. The following are examples of BOX commands with grouping variables specified:

■ **BOX A * B** produces group box plots in which variable A is used to produce a plot for each group defined by variable B.

■ **BOX A * B/GROUPS = 4** produces four box plots using variable A to produce a plot for the first four groups defined by variable B.

To produce box plots for your variables,

1. With the CITY data loaded, type **BOX POP, DENSITY** and press **[Enter]** from the main command screen.

☞ Remember, if you want to print your results, you must type **OUTPUT @** before you conduct the analysis. To direct output back to the screen only, type **OUTPUT ***.

The two box plots shown in Figure 3.9 are produced.

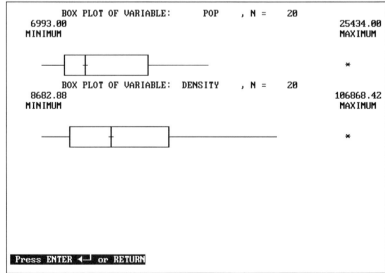

Figure 3.9
Box-and-whiskers plots

Again note that both variables are positively skewed.

☞ When using box plots, both skew and the outside values can be determined, but information may be lost concerning the mode. (The mode is the interval in which most of the scores fall.) However, distributions with two modes often produce large boxes with small whiskers.

■ TPLOT (Series) Graphs

Occasionally you won't have a data set containing values from different people. You may have one that consists of a single subject who has been measured repeatedly. In such a situation, the case number does not constitute a different subject, but a time (a day, a week, or a month). The **TPLOT** command produces plots of a variable against time. These are usually called *series graphs*. It is especially useful for plotting single subjects.

The general form of the TPLOT command is

TPLOT [<VAR>/LAG = <#>, CENTER, NOFILL, STANDARDIZE, MIN = <#>, MAX = <#>]

The LAG option identifies how many cases will be graphed. MYSTAT only plots the first fifteen cases unless you direct otherwise with the LAG option. The CENTER option shades the graph from each point to the center of the distribution. The default is to shade the graph from each point to the left. NOFILL produces a plot without shading. The MIN and MAX options scale the plot. STANDARDIZE converts the variable's raw scores to z scores before plotting the values. A z score is a value from which the variable's mean is subtracted. That result is divided by the standard deviation of the variable. This converts all the values to the same unit of measurement. Using z scores makes comparisons among the values easy. (Z scores are discussed in more detail in Chapter 5.) The following are appropriate TPLOT commands:

- **TPLOT**   plots the first fifteen values of the first numeric variable in the data.

- **TPLOT X/LAG = 10**   plots the first ten values for the variable X.

- **TPLOT X/STANDARDIZE**   plots the standardized values (z scores) of the X variable.

- **TPLOT X/CENTER, MIN = 0, MAX = 20**   plots the first fifteen cases of the X variable. The plot would be shaded to the center of the distribution, and the limits on the plot scale would be set to include values from 0 to 20.

For example, if you were conducting a behavior modification experiment in which you were attempting to decrease a child's hitting behavior, the five-week baseline data (observations in which no intervention occurred) might be in rows 1-5 and the next eight cases might contain weekly observational data after the intervention program was introduced. A typical series graph produced by MYSTAT might look like Figure 3.10.

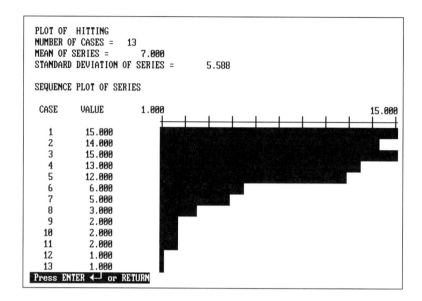

```
PLOT OF HITTING
NUMBER OF CASES =    13
MEAN OF SERIES =       7.000
STANDARD DEVIATION OF SERIES =      5.588

SEQUENCE PLOT OF SERIES

  CASE    VALUE       1.000                                    15.000

    1     15.000
    2     14.000
    3     15.000
    4     13.000
    5     12.000
    6      6.000
    7      5.000
    8      3.000
    9      2.000
   10      2.000
   11      2.000
   12      1.000
   13      1.000
 Press ENTER ⏎ or RETURN
```

Figure 3.10
TPLOT output

To practice producing TPLOTs, perform the following:

1. From the main command screen, type **USE B:PREG** and press
   [Enter] to load the PREG file from the data disk and show the
   variable names.

   This file contains information about human hormone secretions during a typical pregnancy. Each case number represents a week. Ovulation occurs during the second week and
   the child's birth occurs at the fortieth week. The three variables are secretions of estrogens (ESTROGEN), pregnanediol
   (PREGNANE), and chorionic gonadotropin (GONADOTR). The
   amounts of these hormones secreted during a typical pregnancy change dramatically. The hormone secretion values
   have been converted to a common scale for easy comparison.

2. Press **[Enter]** to return to the main command screen.

3. Type **TPLOT ESTROGEN/LAG = 40**, then press **[Enter]** to plot
   the data for estrogen concentrations for all forty weeks.

   MYSTAT will plot the graph shown in Figure 3.11. (Because
   the screen is too small to show the entire graph, you must
   press **[Enter]** to see the rest of the graph.)

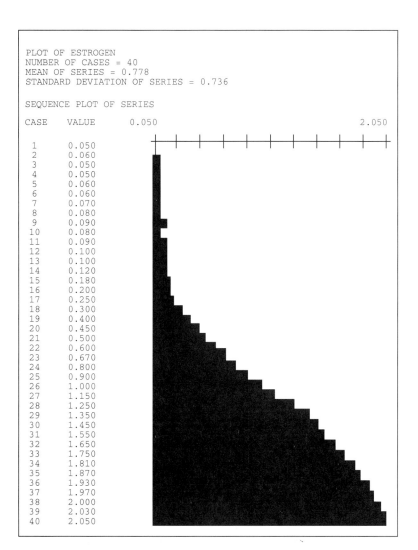

PLOT OF ESTROGEN
NUMBER OF CASES = 40
MEAN OF SERIES = 0.778
STANDARD DEVIATION OF SERIES = 0.736

SEQUENCE PLOT OF SERIES

CASE	VALUE	0.050	2.050
1	0.050		
2	0.060		
3	0.050		
4	0.050		
5	0.060		
6	0.060		
7	0.070		
8	0.080		
9	0.090		
10	0.080		
11	0.090		
12	0.100		
13	0.100		
14	0.120		
15	0.180		
16	0.200		
17	0.250		
18	0.300		
19	0.400		
20	0.450		
21	0.500		
22	0.600		
23	0.670		
24	0.800		
25	0.900		
26	1.000		
27	1.150		
28	1.250		
29	1.350		
30	1.450		
31	1.550		
32	1.650		
33	1.750		
34	1.810		
35	1.870		
36	1.930		
37	1.970		
38	2.000		
39	2.030		
40	2.050		

Figure 3.11
TPLOT of estrogen
secrection during
pregnancy

During a normal menstrual cycle, small amounts of estrogen are secreted. When an egg is fertilized and implants in the uterus, the amount of estrogen produced increases. After four to five months, the placental tissues began to produce large amounts of estrogen. The production of estrogen continues to increase until immediately before birth. The TPLOT graph clearly shows that this increase does not follow a straight line, but increases dramatically from about the thirteenth week to the thirty-second week and then increases more gradually. Just before birth, the estrogen produced is approximately fifty times that produced in a typical monthly cycle.

■ Proper Graphic Design

Graphs should be designed to present information as accurately and concisely as possible. However, the eye can be easily fooled. To make graphs as *truthful* as possible, a frequently cited rule is to make them as simple as possible while conveying the appropriate message. Interested readers will find *The Elements of Graphing Data* by Cleveland (1985) quite useful for designing proper graphs. Huff (1954) illustrates some techniques to tell lies with graphs. The graphs provided by MYSTAT will seldom give viewers interpretation difficulty when the default options are chosen.

———— Exercises

1. Produce and optionally print the histogram, stem-and-leaf diagram, and box-and-whiskers plot for the STATES data set on your data disk. The STATES data contains the following five variables: STATE$ (State abbreviation), POPDEN (the population density), SUMMER (the average summer temperature), WINTER (the average winter temperature), and RAIN (the yearly rainfall). Note the distribution shape for each of these variables and whether each distribution is unimodal or has more than one mode. Also note whether the distribution is symmetrical or skewed.

2. Produce box-and-whiskers plots for the numeric variables in the SCHOOL data using the MDT classification as the grouping variable.

3. Produce and optionally print series graphs (TPLOTs) for the production of chorionic gonadotropin and pregnanediol using the PREG data found on the data disk. Which one of these two hormones is most similar to estrogen in the way it is produced?

4. There has been considerable concern about the increasing rates of infections from the AIDS virus. In the March/April 1991 issue of *The Courier-Africa-Caribbean-Pacific-European Community* journal, Elizabeth White details the numbers of infections over an eight-year period reported to the Caribbean Epidemiology Centre (CAREC). The data are reproduced in the file titled CAREC. Open this file. The first variable is the reporting year (YEAR) from 1982-1989. The second variable (AID_CASE) is the number of reported cases. Produce a series graph (TPLOT) of the number of reported infections. If the graphed trend continues, would you be concerned about the number of reported cases in the future?

5. Also in the same issue of *The Courier*, Simon Horner published data on pages 81-83 detailing the financial assistance given by members of the Development Assistance Committee (DAC). This committee's membership is composed of major countries which have joined together in an attempt to coordinate, adopt, and finance environmental policies that emphasize sustainable development and environmental issues. The file DAC reproduces part of the report. The first variable (COUNTRY$) names the member countries. The second variable (YEAR88) is the amount of money

given by each member country in 1988 in billions of dollars. The third variable (YEAR89) is the number of dollars given in 1989. The fourth variable (GNP89) is the percent of gross national product given in 1989. Using both the 1988 and 1989 figures, are there any countries that can conclude that they are giving enough extra assistance to qualify them as an outside value in a box plot? Which countries, if any are these? The Scandinavian countries appear to be giving higher percentages of their gross national product than others to support sustainable development. Are any of these countries outside values in a box plot?

6. Use the DAC file, and produce both histograms and stem-and-leaf plots for the three quantitative variables. What are the shapes of these distributions? If the shapes change, why do they do so? If you were either a Japanese or a United States DAC member, would you have a preference about which plot would be used to show how your country supports environmentally sustainable projects in developing countries?

7. In your statistics text there are undoubtedly interesting data sets presented in the beginning chapters. Graph some of these variables and note their shape.

4

Descriptive Statistics

In most statistics textbooks, descriptive statistics is covered in two or three chapters usually called "Measures of Central Tendency," "Measures of Dispersion," or "Characteristics of Distributions." The topics covered are modes, medians, means, ranges, percentiles, quartiles, variances, standard deviations, kurtosis, and skewness. The texts use formulas to calculate measures that describe the characteristics of a variable. For example, *means*, *medians*, and *modes* describe the most typical value of a variable. *Ranges*, *quartiles*, *variances*, and *standard deviations* measure the amount of scatter in a set of data values. Calculations are conducted on a population or on a sample from a population. A *population* is defined as all the objects or subjects that are of interest to the researcher. *Samples* are subsets of populations. Calculations conducted on populations are called *parameters* and are symbolized in textbooks with Greek letters. Examples of population parameters are means (μ), variances (σ^2), and standard deviations (σ). Calculations conducted on samples are called *statistics*. Formulas for these sample statistics are derived so that they are *unbiased estimators* of their respective population parameters. A statistic is an unbiased estimator if, on average, its value equals the value of the population parameter. Sample statistics are abbreviated with Roman or English letters. Examples of sample statistics are means ($\overline{X}$), variances (S^2), and standard deviations (S). MYSTAT calculates sample statistics.

Objectives

At the end of this tutorial you should be able to

- Calculate mean, minimum, maximum, sum, standard deviation, variance, skewness, kurtosis, standard error of the mean, and range
- Calculate the above statistics for grouped data
- Print a data set

■ Meanings of the Statistics

Below are brief descriptions of the statistics calculated by MYSTAT and typical formulas for conducting these calculations. Further information is beyond the scope of this tutorial but can readily be found in any introductory statistics text.

1. **Sum**   This is simply found by adding all values together. The Greek symbol Σ simply stands for "add what follows together."

$$\text{sum} = \Sigma X$$

2. **Mean**   This is the simple average for the variable. The n is the sample size.

$$\overline{X} = \frac{\Sigma X}{n}$$

3. **Minimum**   This is the smallest value for the variable.

4. **Maximum**   This is the largest value for the variable.

5. **Variance**   This is a measure of how scattered the data are.

6. **Standard deviation**   This is also a measure of scatter. The standard deviation of a variable is the square root of the variable's variance.

7. **Skew**   Skew measures the symmetry of a distribution. Skew was discussed in detail in Chapter 3 (see Figure 3.3). A perfectly symmetric distribution has a skew equal to zero. If a distribution is positively skewed, the mean is a larger number than the median, and the skew is appreciably greater than zero. If the mean is a smaller number than the median, the distribution is negatively skewed, and the skew is appreciably less than zero. If a distribution is noticeably skewed, it is called *asymmetric*. In a perfectly symmetric distribution, the mean and median will be equal.

8. **Kurtosis**   This is a measure of how *peaked* the distribution is. By definition, normal curves (bell-shaped normal distributions) are considered to be neither peaked nor flat; they are called *mesokurtic*, and their kurtosis values are close to zero. Distributions that come to a peak faster than normal curves are *leptokurtic*, and their kurtosis values are appreciably greater than zero. Distributions that are flatter than normal curves are *platykurtic*, and their kurtosis is appreciably less than zero. The term *mesokurtic* must be memorized, but the other terms can be recalled by remembering that you never want to leap off a leptokurtic curve and that a platykurtic curve can be turned over to serve as a plate. Figure 4.1 illustrates these terms.

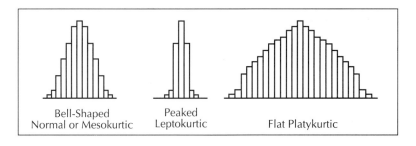

Figure 4.1
Kurtosis of several distributions

Bell-Shaped
Normal or Mesokurtic

Peaked
Leptokurtic

Flat Platykurtic

9. **Standard error of the mean**   This is the standard deviation of the sampling distribution of the mean. This statistic is used as a part of the formula for one-sample *t*-tests, which are discussed in Chapter 5. The standard error of the mean is calculated using the following formula:

$$S_{\overline{X}} = \frac{S}{\sqrt{n}}$$

10. **Range**   This is simply the smallest value recorded subtracted from the largest value recorded. The range gives a very rough indication of the scatter of the scores.

■ Doing the Calculations

Let's calculate descriptive statistics on the ENERGY file stored on the data disk. (For an explanation of the variable names see exercise 5 in Chapter 2.) After the data are loaded, statistics may be calculated on as many numeric variables as you want using the **STATS** command. You may select any combination of statistics available.

The general form of the STATS command is

STATS [<VAR1>, <VAR2>, <...>/KURTOSIS, MEAN, MAX, MIN, RANGE, SD, SEM, SKEWNESS, SUM, VARIANCE BY <VAR3>, <VAR4>, <...>]

Most of the option commands are obvious; for example, MEAN calculates the mean. A few are less obvious. RANGE calculates the variable's range, which is MAX-MIN. SD calculates the standard deviation. SEM calculates the standard error of the mean. SUM calculates the sum or total for the variable values. These options can be listed in any order following the /. If you simply type STATS, only the default statistics minimum, maximum, mean, and standard deviation are calculated.

The variables after the BY option are grouping variables. Descriptive statistics will be produced for each group defined by those variables. To use the BY option, the data must be sorted using these variables and the grouping variables must be numeric. The BY option must follow all other options.
The following are valid STATS commands:

■ **STATS**   produces the default statistics N, MIN, MAX, MEAN, and SD for every numeric variable.

■ **STATS VAR1, VAR2/MEAN**   calculates the mean for the variables VAR1 and VAR2.

■ **STATS VAR1/MEAN BY VAR2**   calculates the mean on VAR1 for each group defined by VAR2.

With MYSTAT loaded and running, the following steps will produce all of the descriptive statistics using the ENERGY data:

1. Load the ENERGY data into MYSTAT by typing **USE B:ENERGY** then pressing [**Enter**]. Press [**Enter**] again to return to the main command screen.

If you wanted to calculate statistics on only one variable, for example, on BTU, you would type **STATS BTU**. If you wanted to do only certain statistics, for example, mean and standard deviation, you would type **STATS/MEAN, SD**. We want to calculate all the statistics on all the variables.

2. Type **STATS/KURTOSIS, MEAN, MAX, MIN, RANGE, SEM, SD, SKEWNESS, SUM, VARIANCE** and press [**Enter**].

You should produce on your screen the output shown in Table 4.1.

TOTAL OBSERVATIONS:	10		
	BTU	SUBSIDY	BTU_DOL
N OF CASES	10	10	10
MINIMUM	0.000	0.600	0.000
MAXIMUM	21.000	15.600	12.556
RANGE	21.000	15.000	12.556
MEAN	8.090	4.390	2.799
VARIANCE	73.997	22.043	15.347
STANDARD DEV	8.602	4.695	3.918
STD. ERROR	2.720	1.485	1.239
SKEWNESS (G1)	0.510	1.476	1.702
KURTOSIS (G2)	-1.444	1.254	1.908
SUM	80.900	43.900	27.988

Table 4.1
Output of the
STATS command

☞ One problem that students encounter when learning statistics is realizing that statisticians sometimes give different things the same name, sometimes give the same thing different names, and don't always use the same abbreviations. MYSTAT gives us two examples. The standard error of the mean was discussed previously. In the output in Table 4.1, MYSTAT titles it STD. ERROR, and in the options it is abbreviated SEM. For standard deviations, texts often use the abbreviation S. In MYSTAT the option for standard deviations is SD, and the output abbreviates standard deviations STANDARD DEV. Finally, MYSTAT adds the labels G1 and G2 to the skewness and kurtosis output.

■ Calculating Statistics for Grouped Data

In data sets, you often have subgroups that are of interest, and you may want to calculate separate descriptive statistics for each group. For MYSTAT to calculate descriptive statistics for groups within a data set, the file must be sorted according to the grouping variables. For example, in the SCHOOL file, the MDT variable defines the group to which each child was assigned. (See Chapter 2, "Reading a Text File," for additional information concerning this data set.) Suppose we want to compare the verbal IQ and performance IQ of each group.

To begin the process you will need to sort the SCHOOL file. Follow these steps:

1. Type **USE B:SCHOOL** and press **[Enter]** twice to load the SCHOOL data into MYSTAT.

   Before you sort data, you need to create a file in which to save the sorted data.

2. Type **SAVE B:SCHLSORT** and press [Enter].

3. Type **SORT MDT** and press [Enter] to sort the file based on MDT and automatically save it to SCHLSORT.

You have just produced and saved a file in which all the cases are sorted (ordered) on the variable MDT. All the children who were assigned to group 1 are first in the file, all the children who were assigned to group 2 are next, and so on.

The following steps will produce separate descriptive statistics for each of the MDT groups:

1. Load SCHLSORT into MYSTAT by typing **USE B:SCHLSORT**, then pressing **[Enter]** twice.

2. Type **STATS VIQ, PIQ/KURTOSIS, MEAN, SD, SKEWNESS, VARIANCE BY MDT**, then press [Enter].

   All the descriptive statistics listed in step 2 for the variables VIQ and PIQ will be calculated separately for every group defined by the MDT variable. MYSTAT presents this information one screen at a time. After each screen of information is presented, MYSTAT pauses and asks you to press **[Enter]** to see the next screen. The analysis for each group has been put into the following seven tables. You should see the output shown in Table 4.2.

Table 4.2a

THE FOLLOWING RESULTS ARE FOR:		
MDT =		1.000
TOTAL OBSERVATIONS:	20	
	VIQ	PIQ
N OF CASES	20	20
MEAN	64.700	69.450
VARIANCE	126.221	136.787
STANDARD DEV	11.235	11.696
SKEWNESS (G1)	-0.117	-0.584
KURTOSIS (G2)	-0.632	-0.366

THE FOLLOWING RESULTS ARE FOR:		
MDT =		2.000
TOTAL OBSERVATIONS:	20	
	VIQ	PIQ
N OF CASES	20	20
MEAN	98.000	94.700
VARIANCE	60.211	131.379
STANDARD DEV	7.760	11.462
SKEWNESS (G1)	0.264	0.324
KURTOSIS (G2)	-0.566	-1.021

Table 4.2b

THE FOLLOWING RESULTS ARE FOR:		
MDT =		3.000
TOTAL OBSERVATIONS:	138	
	VIQ	PIQ
N OF CASES	138	138
MEAN	84.862	90.507
VARIANCE	125.390	166.164
STANDARD DEV	11.198	12.890
SKEWNESS (G1)	0.106	-0.113
KURTOSIS (G2)	-0.086	-0.006

Table 4.2c

THE FOLLOWING RESULTS ARE FOR:		
MDT =		4.000
TOTAL OBSERVATIONS:	8	
	VIQ	PIQ
N OF CASES	8	8
MEAN	102.125	102.250
VARIANCE	57.268	78.500
STANDARD DEV	7.568	8.860
SKEWNESS (G1)	-0.317	-0.502
KURTOSIS (G2)	-0.967	-0.912

Table 4.2d

THE FOLLOWING RESULTS ARE FOR:		
MDT =		5.000
TOTAL OBSERVATIONS:	2	
	VIQ	PIQ
N OF CASES	2	2
MEAN	66.000	88.000
VARIANCE	32.000	18.000
STANDARD DEV	5.657	4.243
SKEWNESS (G1)	0.000	0.000
KURTOSIS (G2)	-2.000	-2.000

Table 4.2e

THE FOLLOWING RESULTS ARE FOR:		
MDT =		6.000
TOTAL OBSERVATIONS:	11	
	VIQ	PIQ
N OF CASES	11	11
MEAN	110.091	115.000
VARIANCE	39.091	136.400
STANDARD DEV	6.252	11.679
SKEWNESS (G1)	1.160	0.256
KURTOSIS (G2)	0.995	-0.973

Table 4.2f

THE FOLLOWING RESULTS ARE FOR:		
MDT =		7.000
TOTAL OBSERVATIONS:	1	
	VIQ	PIQ
N OF CASES	1	1
MEAN	46.000	54.000
VARIANCE	0.000	0.000
STANDARD DEV	0.000	0.000
SKEWNESS (G1)	0.000	0.000
KURTOSIS (G2)	0.000	0.000

Table 4.2g
Descriptive statistics
output for the
SCHOOL data sorted
on MDT

Notice that there are separate outputs for the seven groups within this data set. Look at group 1's output. Their mean IQs are reported as 64.700 and 69.450. Both these scores are within the range for defining a youngster as educable mentally retarded. The children in group 4 had average IQs. These youngsters were not placed in a special educational program. Group 7 contains a single child, so the scatter and shape statistics cannot be calculated. From this analysis, we can see that the groups assigned by the MDT differed dramatically in both VIQ and PIQ. If every value for a variable was the same, which of the descriptive statistics calculated by MYSTAT would have values of zero?

■ Printing Data Sets

It is often easier to edit data files if they can be printed on paper rather than sent to the computer screen. The command OUTPUT @ only prints analyses and graphs. To print data sets, you must use a combination of the USE, OUTPUT, and **LIST** commands.

Try printing the CITY data.

1. From the main command screen, type **USE B:CITY**, then press **[Enter]** twice.

2. Type **OUTPUT @**, then press **[Enter]** to direct the output to the printer and the screen.

3. Type **LIST**, then press **[Enter]**.

   The CITY data is listed on the screen and sent to the printer.

4. Type **OUTPUT ***, then press **[Enter]** to direct subsequent output to the screen only.

——————— Exercises

1. For the data stored in the files CITY, STATES, SCHOOL, and DAC, calculate and optionally print all available descriptive statistics for every quantitative variable. Decide if the variables are symmetrical or skewed. If they are skewed, are they positively or negatively skewed? Describe the kurtosis of each variable. Are the variables mesokurtic, leptokurtic, or platykurtic? Which measure of central tendency (mean, median, or mode) would best describe each variable?

2. For both the STATES and CITY files, you have previously produced stem-and-leaf and box plots. Compare Hspread values and standard deviations using these files. Write a sentence or two describing the relationships, if any, that you found between these statistics.

5

One-Sample Statistical Tests

There are two types of statistics: descriptive and inferential. Up to this point, you have been producing and interpreting descriptive statistics, which are calculated to disclose some characteristic of a sample data set. You have produced measures that indicate typical scores in the distribution (mean, median, and mode), measures of scatter (range, variance, standard deviation), measures of shape (skew and kurtosis), and visualizations of the data values (histograms, stem-and-leaf plots, and box plots).

Inferential statistics are produced to make estimates of population parameters and to make decisions concerning those parameters. The first inferential tests taught in most introductory statistics texts are the one-sample *z*-test and the one-sample *t*-test. You use these tests when you want to determine whether a mean is statistically different from the mean of a known or hypothesized population.

—————— ## Objectives

At the end of this tutorial you should be able to

■ Determine when to conduct a one sample *z*-test or a one sample *t*-test

■ Know how to use the ZCF and ZIF functions within the Data Editor

■ Calculate and interpret one-sample *z*-tests

■ Calculate and interpret one-sample *t*-tests

■ Determining Whether the One-sample *z*- or *t*-Test Is Appropriate

Both one-sample *z*- and *t*-tests are used when the researcher wants to decide if a sample mean is different from a known or hypothesized population mean. If you know the population standard deviation (σ), the most appropriate test is the one-sample *z*-test. If you do not know the population standard deviation, choose the one-sample *t*-test. Figure 5.1 illustrates a decision model that will be expanded in future chapters. The diamonds represent questions, the arrows represent the answers to those questions, and the rectangles depict the decision. Circles are connecting symbols directing you to start or stop or go to another decision diagram in a specific chapter. For example, the circle with the 6 inside it in Figure 5.1 directs you to go to the decision model in Chapter 6.

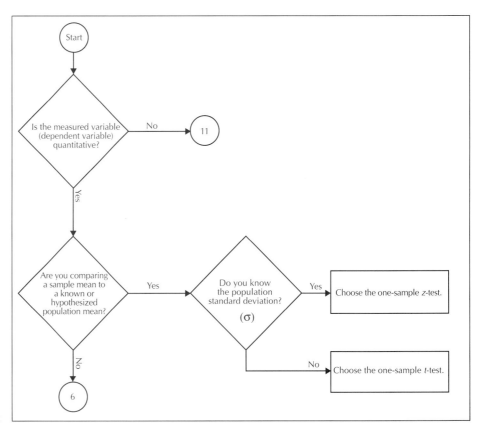

Figure 5.1
Decision model for
one-sample *z*- or *t*-test

If you are conducting a one-sample *z*-test, one of the things you want to know is if the mean of the sample is significantly different from the population mean. To test for significant differences, statisticians set up null and alternative hypotheses. The *null hypothesis* (abbreviated H_0) is a hypothesis used in statistical testing that specifies a value for the population parameter. The *alternative hypothesis* (abbreviated H_1) is a hypothesis that states that the population parameter is something other than that specified by the null hypothesis. These hypotheses can either take directional (one-tailed) or nondirectional (two-tailed) forms. In a *directional* problem, the alternative hypothesis states that the parameter is either greater than or less than some specified value. A *nondirectional* problem's alternative hypothesis simply specifies that the population parameter is not equal to some specific value. The null and alternative hypotheses are always mutually exclusive; that is, if one is true, the other must be false. The null and alternative hypotheses are usually stated in the form of an equation.

Statistical analysis involves hypothesis testing. On the basis of your data analysis, using the appropriate statistical test, you will decide whether you reject or fail to reject the null hypothesis. If you reject a null hypothesis that is correct, you have made an error. This error is called a *Type I error*, and it is determined by setting the *alpha (α) level* or *probability (p)* for an experiment. To decide on the alpha level, you must consider how often you will tolerate making a Type I error. Setting an alpha level of .05 means that there is a five percent probability that a Type I error will occur, or that you will tolerate being wrong when you reject the null hypothesis five percent of the time. While you may choose any alpha, alpha levels of .05, .01, and .001 have been traditionally used by statisticians. Which of these alpha levels you choose depends upon the experimental situation. The smaller the alpha level, the less often you make a Type I error, but a small alpha also makes rejecting the null hypothesis more difficult.

Once you have set the alpha level for a z-test, you need to determine the critical z value that marks the boundary for rejecting the null hypothesis. The *critical value* is the z score, which yields the chosen alpha level in a normal distribution. (For a two-tailed test, divide the alpha level in half.) If your z value is outside the critical value, you reject the null hypothesis; otherwise, you fail to reject it.

Figure 5.2 illustrates that 5% (.05) or fewer of z scores are found below the value of -1.645. The same can be said for a z value of +1.645. For two-tailed tests, 5% of the z values (each tail has 2½%) are outside the values of ±1.96. If the obtained z value is beyond 1.645 for a one-tailed test or beyond either ±1.96 for a two-tailed test, the null hypothesis is rejected. Critical z values are usually found in a table in the appendix of most statistics texts. With MYSTAT, you don't need the table. MYSTAT's ZIF function finds critical z values from alpha levels, and MYSTAT's ZCF function finds the alpha level from observed z values.

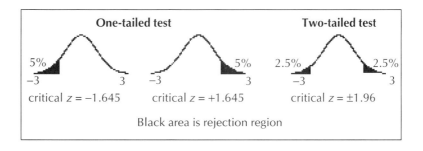

Figure 5.2
Critical z values for
alpha = .05

ZIF

If you enter the alpha level for the experiment into the MYSTAT Editor, the MYSTAT ZIF function will calculate the associated *z* value.

To use the ZIF function to find critical *z* values do the following:

1. At the MYSTAT prompt in the main command screen, type **EDIT**, then press **[Enter]** to start the MYSTAT Editor.

   ☞ If a data set is already loaded in the Edit window, use the NEW Command to clear the Editor.

2. Type **'ALPHA** and press **[Enter]** to name the variable for which you want to enter alpha values.

3. Type **'CRIT_Z** and press **[Enter]** to name the variable for which the critical *z* values will be computed using the ZIF function.

4. Press **[Home]** to place the cursor in the first cell under ALPHA.

5. Type in the alpha values of **.001, .010, .025, .050,** and **.10**.

   ☞ Another way of entering values in a column is to type the value and then press the Down Arrow key instead of **[Enter]**.

The Editor should look like Figure 5.3.

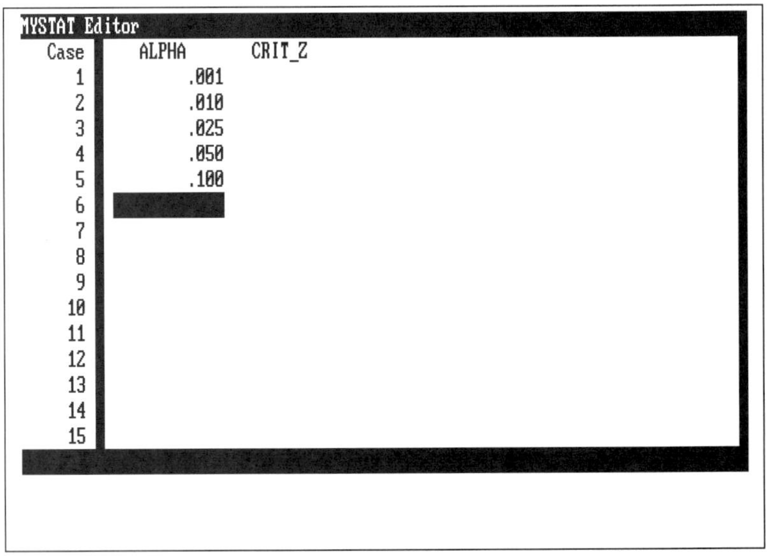

```
MYSTAT Editor
   Case      ALPHA       CRIT_Z
      1         .001
      2         .010
      3         .025
      4         .050
      5         .100
      6     ██████████
      7
      8
      9
     10
     11
     12
     13
     14
     15
```

Figure 5.3
The MYSTAT Editor prepared for the ZIF function

6. Press **[Esc]** to move to the command line.

   You will write a statement so that the variable CRIT_Z is calculated using the ZIF function on ALPHA.

7. Type **LET CRIT_Z = ZIF(ALPHA)**, then press [**Enter**].

MYSTAT calculates the critical *z* values as shown in Figure 5.4.

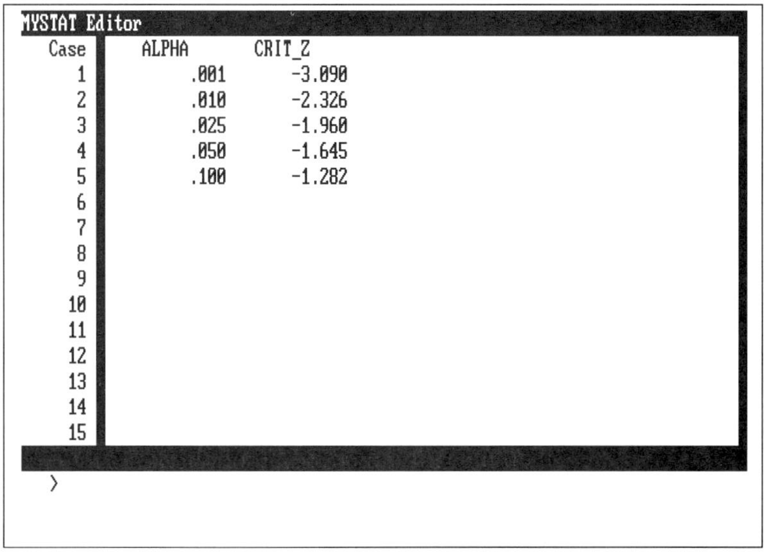

```
MYSTAT Editor
  Case      ALPHA      CRIT_Z
    1         .001      -3.090
    2         .010      -2.326
    3         .025      -1.960
    4         .050      -1.645
    5         .100      -1.282
    6
    7
    8
    9
   10
   11
   12
   13
   14
   15

  >
```

Figure 5.4
Critical *z* values from
the ZIF function

You have just created a table of critical *z* values. Compare the values shown in Figure 5.4 to the ones in your textbook. MYSTAT does the job quickly and can report values that are not contained in the tables because of lack of space. If you want to, save these values as CRITZVAL.

ZCF

The ZCF function is the reverse of the ZIF function. Using MYSTAT and the ZCF function (*z* score cumulative function) you can find the alpha level of any observed *z* value.

Here is how to do it. Suppose you have calculated a *z* value of 2.9457.

1. From the command line, type **NEW** and press [**Enter**] to clear the Editor.

2. Press [**Esc**] to move back to the Edit window.

3. Type **'Z** and press [**Enter**] to name the obtained *z* value.

4. Type **'ALPHA** and press [**Enter**] for the second variable name.

5. Press [**Home**] to move to the first cell under Z, and type **2.9457** for the obtained *z* value.

☞ The Editor will display 2.946 because of the default format.

Your Editor should look like Figure 5.5.

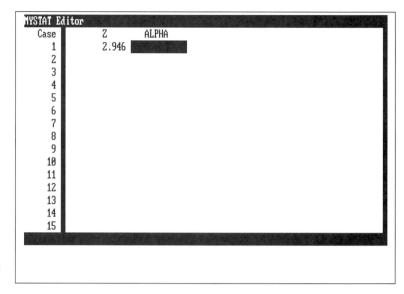

Figure 5.5
MYSTAT Editor with a
z score

6. Press **[Esc]** to move to the command line, then type **LET ALPHA = ZCF(Z)** and press **[Enter]**.

The alpha level will be calculated and the MYSTAT Editor will look like Figure 5.6.

Figure 5.6
Alpha level from the
ZCF function

The alpha calculated is .998. This is the area under a normal curve below a *z* score of 2.9457, or the probability of obtaining a *z* value equal to or below 2.9457. To create the alpha value for a one-tailed *z*-test, create another variable where you subtract .998 from 1.0. To create the alpha level for a two-tailed *z*-test, multiply the one-tailed alpha value by 2. You do not need to save the values you just calculated.

These tables are often found in statistics books. You will find just such a data set on your data disk titled ZVAL. At your leisure, look at it and compare the values found there to the values in your statistics text. (Use the [PgUp], [PgDn], [Home], and [End] keys to scroll through the data.)

■ One-Sample *z*-Test

As noted in Figure 5.1, the one-sample *z*-test is used when you know the population mean (μ) and the population standard deviation (σ) and you wish to compare a sample mean ($\overline{X}$) to the population mean to see if the sample mean is reliably different from the population mean. A *z* statistic is normally distributed. This tutorial will demonstrate how MYSTAT can be used to help do the *z*-test.

The Problem   Suppose that you are a third grade reading teacher. You know the third grade population on average earns a score of 100 with a standard deviation of 15 on a popular reading test. You have decided to introduce a new reading series into your class this year. You wonder if this reading program will change the reading scores. At the end of the academic year you give your students the reading test. Their scores have been entered in the MYSTAT data file titled READ. There are thirty students in this sample. The data are reproduced for you in Table 5.1.

NAME$	READING
Betty B.	100.000
Bob H.	98.000
Claudia W.	89.000
Donald D.	88.000
Donald W.	97.000
Douglas A.	111.000
Dwayne W.	104.000
Eric L.	92.000
Hays D.	134.000
Heidi H.	106.000
Janet T.	99.000
Jean R.	108.000
Jill B.	103.000
John V.	91.000
Judy S.	112.000
Kevin T.	117.000
Lance R.	129.000
Lesley S.	122.000
Linda C.	132.000
Linda G.	107.000

Table 5.1 cont

NAME$	READING
Lisa S.	130.000
Melvin S.	90.000
Michael A.	111.000
Nichole B.	99.000
Pam M.	109.000
Regie T.	118.000
Stanley Q.	120.000
Stephanie D.	102.000
Todd A.	102.000
Todd K.	122.000

Table 5.1
The data from the
READ file

The Solution

Almost all inferential statistical problems can be broken down into six distinct steps. Of these steps, only the third and fourth steps require using the computer. All the others require you to use your head. We'll use the six-step solution to conduct a one-sample z-test.

1. Write the null and alternative hypotheses.

Before you can write the null and alternative hypotheses, you must decide whether you have a directional (one-tailed) or nondirectional (two-tailed) problem. Here the teacher is interested in whether her reading program changes the children's reading scores, so this is a nondirectional (two-tailed) test. The null and alternative hypotheses are

$H_0: \mu = 100$

$H_1: \mu \neq 100$

2. Set the alpha level.

Remember, the alpha level determines how often you will tolerate making a Type I error, and a Type I error occurs when you reject the null hypothesis when it is true. The common alpha level of .05 ($\alpha = .05$) is fine for this example.

3. Collect the data and enter it into MYSTAT.

1. If you are in the MYSTAT Editor, type **QUIT** at the MYSTAT prompt and press **[Enter]**.
2. Load the data into MYSTAT by typing **USE B:READ**, then pressing **[Enter]**.
3. Press **[Enter]** to return to the main command screen from the variable name screen.
4. To see the data, type **LIST** and press **[Enter]**.

   You'll need to press **[Enter]** to scroll through the data, then press **[Enter]** again to get back to the main command screen.

4. Calculate the statistic.

In this case it is the one-sample z statistic whose formula is

$$Z = \frac{\overline{X} - \mu}{\sigma_{\overline{X}}}.$$

μ is the symbol for the *population mean.* Its value is the same as the known or hypothesized mean given in the problem. In this example, the value is 100. The standard deviation of the sampling distribution of the mean is called the *standard error of the mean* and is symbolized by $\sigma_{\overline{X}}$. The value of the standard error of the mean is given by the population standard deviation (σ) divided by the square root of the sample size (n). The formula for the standard error of the mean is

$$\sigma_{\overline{X}} = \frac{\sigma}{\sqrt{n}} = \frac{15}{\sqrt{30}} = 2.7386$$

The only part of the calculation not found directly from the information in the problem is the sample mean, $\overline{X}$. This can be easily found using the computer.

To do this, do the following:

1. From the main command screen, type **STATS/MEAN** and press **[Enter]** to calculate the mean on the numeric variable.

MYSTAT calculates the mean, which is 108.067. Fitting everything into the equation for the one-sample z-test gives an obtained value of

$$Z = \frac{108.067 - 100}{2.7386} = 2.9457.$$

5. Decide whether or not to reject the null hypothesis.

You must either reject or fail to reject the null hypothesis. For z-tests, this decision can be made in two ways. The first is to compare your obtained z value with a critical z value. If your obtained z value is outside of the critical z value, then you reject the null hypothesis. The critical z values for a two-tailed test where $\alpha = .05$ are ±1.96. Your obtained value of 2.9457 is outside 1.96, so you reject the null hypothesis.

The second method is to compare the probability of your obtained z value with the alpha level you set for the experiment. If the probability of your obtained z value is less than the alpha level, you reject the null hypothesis. The probability of a z equal to or outside the value of ±2.9457 is .004. We know this because we calculated this using the ZCF function. Because .004 is less than the alpha of .05, you reject the null hypothesis.

☞ Further information concerning this step can be found in the section "Using ZCF and ZIF."

6. Make a summary statement about the statistical analysis.

In this case you might say something like, "The teacher found that after introducing her new reading procedure, the students' scores significantly changed (Z = 2.946, p < .05)."

■ One-Sample *t*-Test

As noted in Figure 5.1, a one-sample *t*-test is used when a sample mean is to be compared to a known or hypothesized population mean, and the standard deviation of the population is unknown. Since it is likely that we will *not* know the population standard deviation, one-sample *t*-tests are used far more frequently than one-sample *z*-tests. MYSTAT does not have a specific procedure for one-sample *t*-tests but as with the *z*-test, we can work around this and have MYSTAT help us do the calculation.

The Problem

You are a biomedical researcher and you want to know if the consumption of alcohol during pregnancy affects the birth weight of babies. You know that the average birth weight for babies in the United States is 7 lbs., 10 oz. This converts to 7.625 lbs., and is the population mean (μ). You draw a random sample from hospital records of the birth weights of infants whose mothers indicated that they regularly consumed alcohol during pregnancy. The birth weights have been entered in the MYSTAT data file BIRTHWT. Because you do not know the variance or standard deviation of the population, this type of problem is solved using the one-sample *t*-test instead of the one-sample *z*-test.

The Solution

Apply the six-step solution.

1. Write the null and alternative hypotheses.

Because you only want to know if birth weights are changed, this is a two-tailed test. If you wanted to find out whether alcohol consumption decreased birth weight, this would be a one-tailed test.

H_0: μ = 7.625 lbs.

H_1: $\mu \neq 7.625$ lbs.

2. Set the alpha level.

Again select α = .05.

3. Collect the data and enter it into MYSTAT.

The data are already stored for you in the BIRTHWT file.

> 1. Type **EDIT B:BIRTHWT** and press [**Enter**].

> ☞ You could use the **USE** and **LIST** commands instead.

Note that there are thirty-two cases in this file. The variables are the child's first name, NAME$, its birth weight in pounds, WEIGHT, and its sex, SEX$.

4. Calculate the statistic.

You want to calculate the one-sample t statistic. To do this you need to tell MYSTAT what the population mean (μ) is. You will need to create a new variable called POP_MEAN for this data set.

1. With BIRTHWT loaded into the MYSTAT Editor, use the arrow keys to place the cursor in the first empty column in the variable naming row.

2. Type **'POP_MEAN** and press **[Enter]** to name your new variable.

   The hypothesized population mean is 7.625. This is the value that you are going to use for POP_MEAN for every case. This is easy using the LET command.

3. Press **[Esc]** to go to the command line.

4. Type **LET POP_MEAN = 7.625** and press **[Enter]** to set the value of POP_MEAN to 7.625 for every case. Your data should look like Figure 5.7.

```
MYSTAT Editor
    Case       NAME$      WEIGHT       SEX$     POP_MEAN
      1        Stacye      7.100         F         7.625
      2       Monique      6.200         F         7.625
      3        Tracy       5.400         F         7.625
      4        Laura       3.250         F         7.625
      5        Janice      4.980         F         7.625
      6         Lori       9.220         F         7.625
      7        Mahmut      7.140         M         7.625
      8       Richard      5.660         M         7.625
      9       William      6.000         M         7.625
     10      Christine     6.200         F         7.625
     11         Dee        7.250         F         7.625
     12        Diane       7.780         F         7.625
     13        Karen       8.000         F         7.625
     14       Michael      6.660         M         7.625
     15        Julie       4.200         F         7.625

      >
```

Figure 5.7
The BIRTHWT data set

To calculate the t-test, you need to save your new data set, return to the main command screen, and load the data into MYSTAT with the USE command.

5. Type **SAVE B:BIRTHWT** and press **[Enter]**.

6. Type **Y** when asked if this new file can write over the old file.

7. Type **QUIT** and press **[Enter]** to quit the Editor and move to the main command screen.

8. Type **USE B:BIRTHWT** and press **[Enter]** twice to load the data so that statistical analysis can be conducted on the values.

   The **TTEST** command calculates *t*-tests. Its general form is

   TTEST [<VAR1>, <VAR2>, <...>]

9. Type **TTEST WEIGHT, POP_MEAN** and press **[Enter]** to conduct a paired (or dependent) *t*-test on the two named variables.

   The results shown in Table 5.2 should appear.

PAIRED SAMPLES T-TEST ON   WEIGHT   VS POP_MEAN   WITH   32 CASES.

Table 5.2
Results from
completed *t*-test

MEAN DIFFERENCE =   -1.269
SD DIFFERENCE   =   1.288
T   =   5.573   DF =   31   PROB =   .000

These results include a heading or title that indicates the type of *t*-test conducted and the names of the variables used in the calculation. The mean difference between the variable values and the population mean is reported, followed by the standard deviation (SD) of that difference. The *t* value is then reported followed by the *degrees of freedom* (DF). Degrees of freedom are the number of values in the sample data that are free to vary given that you know the value for the statistic. Finally, the statistic, or probability, of obtaining a *t*-statistic whose absolute value is as large or larger by chance, is reported.

5. Decide whether or not to reject the the null hypothesis.

To make this decision for *t*-tests, you look at the probability associated with the *t* value. If the probability of that *t* value occurring by chance is less than the alpha level, then you reject the null hypothesis. Otherwise you fail to reject the null hypothesis. In this case, the probability is 0.000, which is less than the alpha level of .05. Therefore, we obviously reject the null hypothesis that these babies came from a population of infants whose birth weights averaged 7.625 pounds.

WARNING: *If the probability value is less than 0.0005, MYSTAT simply reports that the Prob = 0.000. Statisticians never report that a probability equals 0.000 in a summary statement. You should report zero values as probabilities less than 0.0005.*

6. Make a summary statement about the statistical analysis.

In this case, you might write, "The average birth weight of infants whose mothers consumed alcohol while pregnant significantly differed from the national average (7.625 lbs.) (t = 5.573, df = 31, p < 0.0005)."

Exercises

1. It is known that for a population of joggers the mean time required to run one mile is 10.2 minutes with a standard deviation of 1.16 minutes. A sample of joggers is given a special fitness training program and then tested on the time they take to finish a one-mile run. The data are found in the file JOGGERS. Did the fitness program significantly affect the joggers' times? Write the null and alternative hypotheses along with your summary statement.

2. It is hypothesized that the drug tamoxifan significantly increases the survival time of women who have been diagnosed as having a certain type of aggressive breast cancer. It is also known that, on average, women with this cancer live only 2.75 years after diagnosis. The Women's Cancer Center has given a random sample of patients tamoxifan and notes their survival rates over a five-year span. State the null and alternative hypotheses. The data are contained in the data set titled CANCER. Did tamoxifan increase these women's survival time? Write the null and alternative hypotheses along with your summary statement.

3. Use the STATES data set. Determine whether the average rainfall in the United States in the year surveyed was 20 inches. Write the null and alternative hypotheses along with your summary statement.

4. Answer any assigned problems from your text using MYSTAT. Write the null and alternative hypotheses along with your summary statement.

6

Two-Sample Statistical Tests

Sometimes you will need to compare the means of two samples to determine whether they are different. If a treatment is given to one sample and not to the other and the means of the samples are different, the treatment may have produced this difference. In this chapter, we investigate differences between the means of two samples. This is the classic application of the *t*-test.

Two *t*-tests will be discussed in this chapter: the *independent* t-*test* and the *dependent* t-*test*. In some texts the independent *t*-test is called the two-sample *t*-test. The dependent *t* might be called the paired or correlated *t*-test. The dependent *t*-test also may be called a two-sample test when subjects are matched on an important variable like age or weight.

Objectives

At the end of this tutorial you should be able to

- Determine when to use a dependent or independent *t*-test
- Use MYSTAT to calculate dependent and independent *t* values
- Interpret *t*-test results
- Know the assumptions underlying these tests

■ Independent Versus Dependent *t*-Test

Both the independent and dependent *t*-tests are used to decide whether two sample means are different from each other. If the two samples are not related to each other, conduct an independent *t*-test. If there is a relationship between the two samples, use the dependent *t*-test. Figure 6.1 illustrates this decision model. Remember that circles are connecting symbols directing you to start or stop or go to another decision diagram in a specific chapter. For example, the circle with the 5 inside it tells you that this decision diagram continues from Chapter 5.

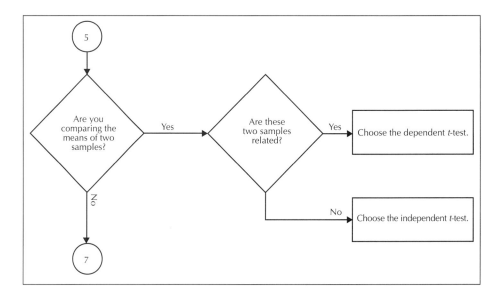

Figure 6.1
Decision model for independent or dependent *t*-test

■ The TTEST Command

The TTEST command you learned in Chapter 5 needs to be expanded for two sample tests. The complete, general form of the TTEST command is

TTEST [<VAR1>, <VAR2>, <...> [* GROUP]]

This command is used to conduct both independent and dependent (paired) *t*-tests. If commas are used between the variables, dependent *t*-tests are conducted. If an asterisk separates variables, then independent *t*-tests are conducted. The following are valid TTEST commands:

- **TTEST A, B**   conducts a dependent *t*-test between variable A and variable B.

- **TTEST A, B, C**   conducts a dependent *t*-test between each pair of variables. Thus, a *t*-test would be conducted between A and B, another *t*-test between A and C, and a third *t*-test between variables B and C.

- **TTEST A * GROUP**   conducts an independent *t*-test using the A variable and the GROUP variable. The TTEST command compares the difference between the means of the A variable defined by the GROUP variable. (Note that the variable named GROUP can only contain two different values because independent *t*-tests only are appropriate for two groups.)
- **TTEST A, B, C * TYPE**   conducts three separate independent *t*-tests. The first is between the A variable using the TYPE variable to define the groups. The second *t*-test is between B using TYPE, and the final independent *t*-test is between C using TYPE. (Again, TYPE can only have two different values.)

■ Independent *t*-Test

As noted in Figure 6.1, the *independent* t-*test* is used when the two samples are unrelated (the people in the first sample are unrelated to the people in the second sample). This is the case when subjects are randomly assigned to an experimental group or a control group. We will illustrate a problem using MYSTAT and the six-step solution.

The Problem

You are the project director for a pharmaceutical company that is marketing a new drug for the treatment of the HIV virus. You have gone through the toxicology tests and you are now working on an experiment with monkeys. You randomly select and assign fifty subjects that are HIV positive to two different groups. The control group gets a pill with no active ingredients (a placebo). The treatment group receives your drug. To test the effectiveness of the drug, your company has devised a sophisticated method to measure the degree to which a monkey's immune system has been compromised. The numbers run from zero (no compromise) to 100 (the immune system is fully compromised). After one year both groups are examined and their immune system scores are tabulated. You want to see whether the two groups' immune systems have responded differently.

The Solution

Below is the six-step solution.

1. Write the null and alternative hypotheses.

The null hypothesis is that there are no differences between the two samples. If the null hypothesis is true, then the two samples come from populations that have identical means (the population parameters are identical).

$H_0: \mu1 = \mu2$

$H_1: \mu1 \neq \mu2$

2. Set the alpha level.

For this experiment set the α level to .05.

3. Collect the data.

These data are provided in the file AIDS. Note when you use this file that there is a variable labeled GROUP. This is the group to which each subject was assigned. (The subjects given the number 1 are in the control group; the subjects given the number 2 are in the experimental group.) The next variable, IMSCORE, is the subject's immune system score. There are twenty-five group 1 subjects and twenty-five group 2 subjects.

4. Calculate the statistic.

1. From the main command screen, type **USE B:AIDS**, then press [**Enter**] to load the AIDS file into MYSTAT.

2. After viewing the variable names, press [**Enter**] again to return to the main command screen.

3. Type **TTEST IMSCORE * GROUP**.

   IMSCORE is the dependent variable in the analysis. The dependent variable is the variable that is measured in the experiment and the one whose mean is compared across the two sample groups. The GROUP variable indicates to which of the two sample groups the dependent variable (IMSCORE) belongs. The asterisk before GROUP indicates that GROUP is the grouping variable.

4. Press [**Enter**] to conduct the *t*-test.

   The results are presented in Table 6.1.

INDEPENDENT SAMPLES T-TEST ON IMSCORE GROUPED BY GROUP

GROUP	N	MEAN	SD
1.000	25	32.760	21.008
2.000	25	18.240	10.760

Table 6.1
Independent
t-test results

SEPARATE VARIANCES T = 3.076 DF = 35.8 PROB = .004
POOLED VARIANCES   T = 3.076 DF = 48   PROB = .003

We assume that not only are μ1 and μ2 equal (see the null hypothesis in step 1 of the six-step solution) but that the variances of the two populations are equal. To estimate that common variance, we combine (or pool) the variances from both samples, and MYSTAT reports this as a *pooled variance*. Both a separate variances and a pooled variances *t* value are reported. The separate variances *t* value used the separate variances within each group for its calculation. If variances differ substantially, use the separate variances *t* value. See exercise 2 at the end of this chapter for practice on this.

The degrees of freedom (DF) for this problem are also reported. Remember that degrees of freedom were defined in Chapter 5 as the number of values in the sample data that are free to vary given that the value for the descriptive statistic is known. The separate variances degrees of freedom is calculated differently; they decrease as the variances become more different. MYSTAT always provides the degrees of freedom when it conducts a statistical test that uses them.

5. Decide whether or not to reject the null hypothesis.

The probability of .003 is smaller than the alpha level of .05 set in step 2; therefore, you reject the null hypothesis.

☞ The decision-making process for rejecting the null hypothesis when using the *t*-test is the same as that used for the *z*-test. If the probability is less than the alpha level, reject the null hypothesis.

6. Write a summary statement.

Since you rejected the null hypothesis, the summary statement might go something like this: "There was a significant difference between the immune system scores for the subjects who took the new drug when compared to those who received no treatment. Subjects in the treatment group had immune systems that were less compromised than those in the control group (t = 3.076, df = 48, p = .003, two-tailed)."

■ Dependent *t*-Test

The *dependent* t-test is used when the subjects in the experiment are measured twice or matched on one or more attributes. Dependent *t*-tests can be used for both one and two-sample situations. The same statistical formula is used for both.

Two-Sample Situation

The dependent *t*-test is used with two separate groups when the cases or subjects are matched. This typically happens when the investigator matches subjects (makes sure they are the same in some way) and randomly assigns one member of the matched pair to the experimental group and the other member to the control group. This design frequently is employed in educational research.

The Problem You are the psychologist responsible for program evaluation in a community-based educational facility for severely retarded children. One of the programs attempts to teach these children self-care skills. The literature is not clear whether imitation learning or physically directed learning is the best method for teaching these skills. In imitation learning the child watches the teacher complete a task and then is asked to repeat the task by remembering what the teacher did. In physically directed learning, the teacher actually physically guides the child through the process. You have directed the staff to place half the children in an imitation learning curriculum for three months and the other half in a physically directed curriculum. Previous research has shown that the learning rates for these children depend upon their intellectual level. Therefore, each subject is matched on intellectual level and one member from each pair is randomly assigned to the imitation learning condition. The other member from the pair is assigned to the physically directed group.

You are interested in detecting whether the physically directed method is better than the imitation method, which makes it a one-tailed or directional test. The data collected are the number of skills learned under each method.

The Solution

The now familiar six-step solution is easily used to solve the problem.

1. Write the null and alternative hypotheses.

Not all textbooks follow the same rules for directional hypotheses. This text adopts a specific way to set up these hypotheses. In this case you expect to find that physically directed learning is better (produces higher scores) than imitation learning. Therefore the alternative hypothesis will indicate that the mean of the second group (imitation learning) will be larger than the mean of the first (physically directed learning). The null hypothesis must include all other options (i.e., that the means of the groups are equal and that the mean of group 2 will be smaller than the mean of group 1). Below are these two hypotheses:

H_o: $\mu 1 \geq \mu 2$

H_1: $\mu 1 < \mu 2$

2. Set the alpha level.

We'll set the alpha level to .05.

3. Collect the data.

The data for this project are in a file titled SELFCARE which contains imitation scores and physical guidance scores for twenty-four pairs of children.

4. Calculate the statistic.

The formula for the dependent *t* is

$$t = \frac{\overline{D}}{S_{\overline{D}}}$$

D is the difference between the scores for each matched pair. In the numerator, $\overline{D}$ is the the average of those differences. In the denominator $S_{\overline{D}}$ is the sample standard deviation of the mean differences.

To obtain the mean difference value ($\overline{D}$), MYSTAT always subtracts the second variable listed in the command from the first. Note that the sign of $\overline{D}$ is affected by the order in which these variables are chosen. It is a good idea to select the variables in the same order as they are written in the alternative hypothesis so that the sign of the mean difference corresponds to the alternative hypothesis. If the sign in the numerator changes, the sign of the *t* value changes.

To calculate the *t* value,

1. Type **USE B:SELFCARE**, in the main command screen, then press [**Enter**].

2. After viewing the variable names, press [**Enter**] again to return to the main command screen.

3. Type **TTEST IMMITATI, PHYSICAL**.

   MYSTAT will subtract the values of the PHYSICAL variable from the values of the IMMITATI variable to compute the difference scores.

4. Press [**Enter**] to conduct the *t*-test.

   You will obtain the results shown in Table 6.2.

Table 6.2
Results from the
paired *t*-test

PAIRED SAMPLES T-TEST ON IMMITATI VS PHYSICAL WITH 24 CASES

MEAN DIFFERENCE =   -3.250
SD DIFFERENCE   =   3.674
T =   4.333   DF = 23   PROB = .000

5.   Decide whether or not to reject the null hypothesis.

MYSTAT always reports the probabilities or Type I errors for two-tailed tests. Thus, for directional tests you must divide the reported probability in half. Then you must check that the mean difference reported occurs in the direction expected by the alternative hypothesis. If the mean difference is negative, then the *t* value is negative. MYSTAT, however, always reports the *t* value as a positive number. Therefore, when reporting your analysis, you must change the sign of the *t* value to match the sign of the mean difference. The *t* value of -4.333 has a low probability of occurring by chance. Remember that 0.000 is output if the probability is less than 0.0005. This value is lower than the alpha set in step 2, so you reject the null hypothesis.

6.   Write a summary statement.

The summary statement for this research could be written: "The author found that when teaching self-care skills to severely retarded children, a physically-directed method produced significantly better results than an imitation method (t = -4.333, df = 23, p < 0.0005, one-tailed)."

One-Sample Situation

In one-sample designs, the dependent *t*-test formula is used when the same people are measured twice. Another name for the one-sample dependent *t*-test is the *paired* t-*test*. This repeated-measures situation occurs when the same subjects are exposed to two different treatments or when pretest and posttest scores are used. The same people are the subjects in both conditions. A one-sample situation requiring a paired *t*-test occurs when the investigator pretests subjects, gives them a treatment designed to change their scores, and then posttests them.

If you are interested in detecting a difference between scores, a nondirectional (two-tailed) approach is appropriate. If you are interested in detecting whether there was an increase (or decrease) in the scores between variables or a difference in the pretest and posttest scores, then a directional (one-tailed) test is needed.

The Problem

You are a biologist interested in the effects of alcohol consumption on maze-running performance in hamsters. You have trained ten hamsters to run a maze until their times show no improvement. After running the maze again as a pretest measure, the rats are given alcohol so that their blood-alcohol levels are proportional to the level that makes humans legally intoxicated. The rats then run the maze a second time, and the results are measured.

You want to detect whether the pretest and posttest times differ. This is a nondirectional (or two-tailed) test. The dependency here is quite evident, as the same rats are being measured twice. The paired *t*-test formula is used.

The Solution

The six -step solution follows.

1. Write the null and alternative hypotheses.

Below are the two hypotheses:

H_o: $\mu 1 = \mu 2$

H_1: $\mu 1 \neq \mu 2$

The null hypothesis indicates that the pretest scores (time to run the maze before alcohol) are equal to the posttest scores.

2. Set the alpha level.

We'll use .05 again.

3. Collect the data.

The data for this project are so few that it might be valuable to review keyboard entry by typing in this data set.

1. Type **EDIT** and press **[Enter]** to get to the MYSTAT Editor.

☞ Remember to use NEW to clear the Edit window if necessary.

2. Enter the data shown in Figure 6.2.
3. Press **[Esc]** to get to the command line.

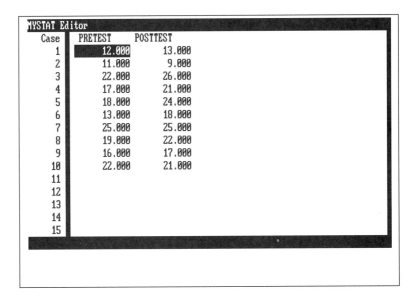

Figure 6.2
The HAMSTERS data

Now you need to save the data so that you can conduct statistical analyses.

4. At the MYSTAT prompt, type **SAVE B:HAMSTERS**, then press **[Enter]**.

5. Type **QUIT** and press **[Enter]** to return to the main command screen.

4. Calculate the statistic.

1. Type **USE B:HAMSTERS** and press **[Enter]** twice to load the data so that statistical analyses can be conducted on the values.

2. Type **TTEST PRETEST, POSTTEST** and press **[Enter]**.

   You should obtain the results shown in Table 6.3.

PAIRED SAMPLES T-TEST ON PRETEST VS POSTTEST WITH 10 CASES

Table 6.3
Results of the
paired *t*-test for the
HAMSTERS data

MEAN DIFFERENCE = -2.100
SD DIFFERENCE   =   2.685
T =   2.473   DF = 9   PROB = .035

5. Decide whether or not to reject the null hypothesis.

Note that the mean difference is negative, so the *t* value must be negative. The *t* value of -2.473 has a lower probability of occurring by chance (.035) than the alpha you set at step 2 (.05). Reject the null hypothesis.

6. Write a summary statement.

The summary statement could be written: "The author found that alcohol consumption in hamsters significantly changed (increased) the time it took them to run the maze (t = -2.473, df = 9, p = .035)."

———————— Exercises

1. A large group of learning-disabled college freshmen who experience debilitating anxiety before major tests were matched on an index of test anxiety. Members of these matched pairs were randomly assigned to different groups. The first group was given two weeks of relaxation exercises (RELAX). The second group of students was given two weeks of study skills training (STUDY). Using the data found in the file ANXIETY, determine whether there is a significant difference between the final exam scores of the two groups. Hand in the printed output if your instructor requests it along with your final summary statement.

2. A vicious debate between behaviorists and traditional medical doctors concerns the merits of using the stimulant Ritalin in treating childhood hyperactivity. A large group of hyperactive children in an urban school system were randomly assigned to either a behavior modification program or a drug therapy program. The data are contained in the file HYPER. Determine if there is a significant difference on these children's out-of-seat behavior (OUTSEAT) between the two methods.

3. Divide the STATES data set into two groups (you will need to create a new variable for these groups). The first group is those states east of the Mississippi River. The second group is those states west of the Mississippi. Are there significant differences between their population densities?

4. Using the grouping variable created in exercise 3, are there significant differences between their summer temperatures, winter temperatures, and rainfall amounts? Write summary statements for each.

5. The attitudes toward obtaining an advanced college degree of fifteen undergraduate minority students were measured before and after they participated in a federally funded program designed to increase their awareness of the benefits of a higher degree. The higher the score, the more positive the subject's attitude. The data are contained in the file HDEGREE. Did attending the program significantly change their attitudes?

6. Using the DAC file, determine whether there has been a change in spending by DAC committee members from 1988 to 1989. Write a summary statement.

Analysis of Variance

Two sample *t*-tests determine if there are differences in mean scores between two groups. Analysis of Variance (ANOVA) procedures determine differences among means when there are more than two groups. In ANOVA, grouping variables are referred to as *factors*. If there is a single grouping variable, the procedure is called one-way ANOVA. If there are two grouping variables, the procedure is called two-way ANOVA. With two or more grouping variables, the procedure is called multi-way ANOVA.

Grouping variables or factors are also called *independent variables* in research. In true experiments, researchers randomly assign subjects to the different groups. Researchers control the independent variables (they decide who goes into which treatment group). They want to detect whether some measured variable has different values depending upon the group. This measured variable is called the *dependent* variable.

ANOVA is used instead of multiple *t*-tests to reduce the probability of making a Type I error. If you assign subjects to three groups and want to determine if there are differences in the dependent variable among these three groups, using only what you have previously learned, you could conduct a *t*-test comparing group 1 and group 2, then a *t*-test comparing group 1 and group 3, and finally, a third *t*-test comparing group 2 and group 3. However, there is a serious problem with this procedure. If each of the three *t*-tests had alpha levels (Type I errors) set at .05, the Type I error increases for the experiment when you do three tests in a row. Using ANOVA procedures avoids this Type I error inflation.

Objectives

At the end of this tutorial you should be able to

■ Determine whether you have a one-way or multi-way ANOVA
■ Calculate one-way ANOVA results using MYSTAT
■ Calculate two-way ANOVA results using MYSTAT

■ One-Way ANOVA Versus Multi-Way ANOVA

Both one-way and multi-way ANOVA procedures are used to decide if three or more sample means are different from one another. To choose between these procedures, determine whether the groups are defined by one or more grouping variables. Figure 7.1 illustrates the decision model for determining if you should choose a one-way or multi-way ANOVA.

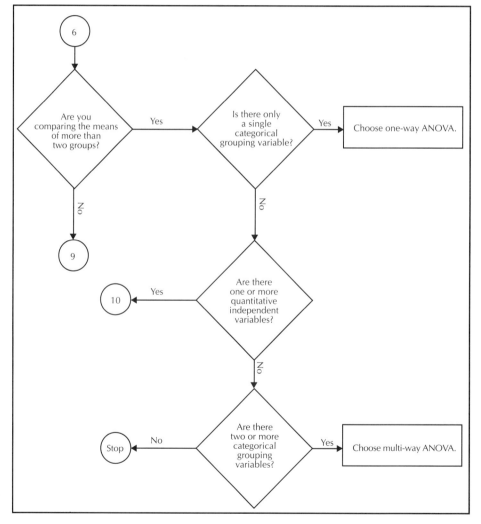

Figure 7.1
Decision model for
one-way or multi-way
ANOVA

■ The ANOVA Commands

To conduct an analysis of variance using MYSTAT, three separate commands must be used. The first command is the CATEGORY command, which defines the variable(s) that identify the independent variable(s), also called the categorical variable or the factor and indicates the number of levels (groups) contained in each factor.

☞ The categorical variable must be coded with successive integers.

The second command is the ANOVA command. The ANOVA command identifies the dependent variable. The third command is the ESTIMATE command, which tells MYSTAT to do the calculations.

For example, suppose a researcher is interested in detecting whether income differs across the four groups. The categorical variable GROUP has four levels, and the dependent variable is INCOME. After loading the data into MYSTAT, the following commands would conduct a one-way ANOVA:

>   **CATEGORY GROUP = 4, [Enter]**
>   **ANOVA INCOME, [Enter]**
>   **ESTIMATE, [Enter]**

For data where there are two levels for factor A, three levels for factor B. and the dependent variable is Y, the following commands would conduct a two-way ANOVA:

>   **CATEGORY A = 2, B = 3, [Enter]**
>   **ANOVA Y, [Enter]**
>   **ESTIMATE, [Enter]**

Researchers refer to this as a 2 x 3 (2 by 3) ANOVA.

■ One-Way ANOVA

Figure 7.1 explains that one-way ANOVA is used to detect mean differences among groups when there is a single independent variable and the number of groups formed by that independent (grouping) variable is three or more. One-way ANOVAs are found frequently in the research literature and are one of the most important analysis tools you can learn.

The Problem   You are responsible for determining how different newspaper coupons for your company's product affect sales. You set up four groups. For group 1, a newspaper advertisement without a redemption coupon will be printed. For group 2, the same advertisement will be used with a coupon worth 10 cents. For group 3, the advertisement and a 30-cents coupon will be used. For group 4, the advertisement and a 50-cents coupon will be used. You select twenty communities with daily newspapers within your sales region and randomly assign these communities to your four groups. For each group the local newspaper carries an advertisement for your product in the Wednesday edition.

Sales figures from the twenty communities are totaled on the following Saturday (the coupon's expiration date).

The Solution

Below is the six-step solution.

1. Write the null and alternative hypotheses.

In ANOVA, the null and alternative hypotheses are stated somewhat differently. The null hypothesis always assumes that all the groups come from populations with identical means. In this case, the null hypothesis indicates that average sales of your product are the same regardless of the coupon value in the newspaper. The alternative hypothesis is simply that any of the group means differs. Because there are so many possibilities for the alternative hypothesis the null and alternative hypotheses are usually written

H_0: $\mu 1 = \mu 2 = \mu 3 = \mu 4$

H_1: not H_0

2. Set the alpha level.

We'll set the alpha level to .05.

3. Collect the data.

The data for this problem are found in the file ADVERT.

1. Type **USE B:ADVERT** and press **[Enter]** from the main command screen. Notice that there is a variable for the number of products sold (SOLD) and the group to which the community was assigned (GROUP).

2. Press **[Enter]** to return to the main command screen.

4. Calculate the statistic.

In this study, the dependent variable is the number of products sold (SOLD), and the independent variable is the four treatments (GROUP).

1. Type **CATEGORY GROUP = 4** and press **[Enter]** to define the independent variable as the variable named GROUP and indicate that the number of levels contained in this variable is 4.

2. Type **ANOVA SOLD** and press **[Enter]** to indicate that the dependent variable is SOLD.

3. Type **ESTIMATE** and press **[Enter]** to direct MYSTAT to begin the calculations.

The results you should receive are shown in Table 7.1.[1]

DEP VAR: SOLD   N: 20   MULTIPLE R: .525   SQUARED MULTIPLE R: .275

ANALYSIS OF VARIANCE

Table 7.1
One-way
ANOVA results

SOURCE	SUM-OF-SQUARES	DF	MEAN-SQUARE	F-RATIO	P
GROUP	15940.000	3	5313.333	2.026	0.151
ERROR	41960.000	16	2622.500		

In one-way ANOVA two degrees of freedom (DF) are calculated. The first degree of freedom reported is associated with the number of groups in the analysis (number of groups – 1). The second is associated with the error term (number of cases – number of groups). In some texts the group degrees of freedom is called the degrees of freedom in the numerator or the degrees of freedom between groups. The degrees of freedom due to error is called the degrees of freedom in the denominator or the degrees of freedom within groups. In every one-way ANOVA, MYSTAT calculates these values for you.

5. Decide whether or not to reject the null hypothesis.

Look at the MYSTAT probability value. Since the *p* value of 0.151 is larger than the alpha of .05, you fail to reject the null hypothesis.

6. Write a summary statement.

In this case the researcher reports, "Enclosing coupons with the advertisement did not create a significant difference in the sales of the product (F = 2.026, df = 3, 16, p = 0.151)." Remember that the 3 degrees of freedom are used to calculate the mean square *between* groups and the 16 degrees of freedom are used to calculate the mean square *within* groups.

■ Two-Way ANOVA

Experiments with more than one grouping variable are called factorial designs. Factorial designs are labeled by the number of levels of each factor (grouping variable). For example, assume that you are conducting an experiment to measure the blood pressure of men and women while at rest and while pedaling a stationary bicycle. This is a 2 x 2 factorial design. The first factor, SEX, has two levels (Male and Female), and the second factor, EXERCISE, also has two levels (Resting and Pedaling).

There are three things to be measured in this experiment:

1. Does male blood pressure differ from female blood pressure? The test of the SEX main effect answers this question. It is like doing a one-way ANOVA on SEX.

2. Do the blood pressures of people at rest differ from their blood pressures while exercising? The test of the EXERCISE main effect answers this question. It is like doing a one-way ANOVA on EXERCISE.

3. Is the difference between male blood pressure and female blood pressure while resting the same as the difference between their blood pressures

while exercising? The test of the interaction between SEX and EXERCISE answers this question. A significant interaction means that differences in resting blood pressures and exercise blood pressures must be described separately for males and females.

The two-way ANOVA procedure addresses all three of these questions.

☞　There isn't any theoretical upper limit to the number of factors or the number of levels in ANOVA designs. However, because the number of tests increases with increases in the independent variables, the results become more difficult to interpret. With every new variable you also need more subjects to create the groups. In business and social sciences, most designs involve two or three factors with four or fewer levels.

The Problem

You are responsible for testing the effectiveness of a drug combined with therapy on schizophrenic patients' behavior. The drug has three dosages (absent, low dosage, high dosage) and the therapy has four types (behavior modification, psychodynamic, group counseling, nondirective). This is referred to as 3 x 4 factorial design because the first factor has three levels and the second factor has four levels. Assume that you have 120 subjects who are randomly assigned to the twelve groups; each group contains ten patients.

The Solution

Work the drug dosage and therapy problem with MYSTAT using the six-step solution.

1. Write the null and alternative hypotheses.

In two-way ANOVA analyses, there are three sets of null and alternative hypotheses: one for each factor, and one for the interaction. For main effects, there are no differences between the means or there are no differences in treatment effects for that factor. The alternative hypotheses are simply that the null is false. For the interaction hypothesis the null is simply that there is no interaction between the two variables. Again, the alternative is that the null is false. In this tutorial we will state two-way ANOVA hypotheses without symbols.

H_{01}: The means for the drug dosage groups are all equal.

H_{11}: H_{01} is false.

H_{02}: The means for the therapies are all equal.

H_{12}: H_{02} is false.

H_{03}: There is no drug dosage by therapy type interaction.

H_{13}: H_{03} is false.

2. Set the alpha level.

We'll use .05 again for the alpha level.

3. Collect the data.

The data for this problem are found in the file SCHIZOPH.

1. From the main command screen type **USE B:SCHIZOPH** and press [**Enter**].

   In the variable name screen, note that there is a variable for the level of drug, DOSAGE, a variable for the TYPE of therapy, and a variable for the number of behavioral INCIDENTs the person had during the experiment. The fewer behavioral incidents the better the recorded behavior for the individual.

2. Press [**Enter**] to return to the main command screen.

4. Calculate the statistic.

You want to see the effects on INCIDENT due to DOSAGE and TYPE.

1. Type **CATEGORY DOSAGE = 3, TYPE = 4** and press [**Enter**] to set up two independent variables, one with three levels and one with four levels.

2. Type **ANOVA INCIDENT** and press [**Enter**] to indicate that INCIDENT is the dependent variable.

3. Type **ESTIMATE** and press [**Enter**] to calculate the statistic. (If you are using an older, slower PC, this may take up to two minutes.)

   The output shown in Table 7.2[2] should appear.

DEP VAR: INCIDENT   N: 120   MULTIPLE R: .360   SQUARED MULTIPLE R: .130

ANALYSIS OF VARIANCE

SOURCE	SUM-OF-SQUARES	DF	MEAN-SQUARE	F-RATIO	P
DOSAGE	2.867	2	1.433	3.448	0.035
TYPE	1.158	3	0.386	0.929	0.430
DOSAGE* TYPE	2.667	6	0.444	1.069	0.386
ERROR	44.900	108	0.416		

Table 7.2 Two-way ANOVA results

5. Decide whether or not to reject the null hypothesis.

In this case you reject the first null hypothesis (H_{o_1}) and fail to reject the second and third nulls. Again, all you need to do is look at the *p* values and check them against your chosen alpha level.

6. Write a summary statement.

For this problem, the researcher might report: "The dosage of the drug administered to the schizophrenics made a difference in the number of behavioral incidents subjects displayed during the experimental period (F = 3.448, df = 2, 108, p = 0.035). No differences due to the type of therapy were detected (F = 0.929, df = 3, 108, p = 0.43). Finally, there was no evidence of an interaction between the dosage level of the drug and the individual's type of therapy (F = 1.069, df = 6, 108, p = 0.386)."

——————— Exercises

1. Use the STATES data set. Divide the states into three regions (you will need to make a variable to code these different regions). The three regions are the East, the Midwest, and the West. Using these groups, are there significant differences in population density, summer temperature, winter temperature, and rainfall? Write summary statements for each of the problems. Your answers may differ from others depending on how you divided the states into the groups.

2. An experimental psychologist is investigating the effect on rats of delta-9 tetrahydrocannabinol (THC) ingestion and their maze-running ability. THC is the psychoactive ingredient in marijuana. The researcher randomly assigns ten rats to four different treatment levels (None = 1, Low = 2, Medium = 3, and High = 3) of THC. After one hour she places each rat into a maze. The time it takes each of the 40 rats to complete the maze (MAZETIME) is then recorded. The data is found in the file RATS. Is there a significant effect on maze-running ability across the THC levels? Write an appropriate summary statement.

3. Travelers from the United States have frequently been warned about eating the foods in foreign countries. The *Herald* magazine in May 1991 reported data from a survey of food available from street venders in Karachi, Pakistan. The survey systematically sampled foods across several different areas in the city. Partial data from that survey are contained in the file KARACHI. The variables are defined as follows: The type of food (a drink or snack) is labeled FOOD_TY$. The name of the food is labeled FOOD_NA$. The area from which the food was sampled is called AREA. There were eight areas sampled. The bacterial count in the food is called BACTERIA. This variable is measured in colony-forming units per gram of food in powers of 10. Thus if there were 100 colony forming units per gram, a score of 2 (10^2) would be recorded. Finally, the presence of LEAD, ARSENIC, CADMIUM, and CYANIDE are recorded (1 = present, 0 = absent). The report found that only 15.73% of the food sampled in the study was perfectly safe to consume. The safe food was almost exclusively bottled soft drinks. These are referred to in the file as colas. The full report found that 16% of the food was acceptable to eat, 29.21% if eaten would put the person at risk, and 38.95% of the food was dangerous to consume. Use the KARACHI file and determine if there is a significant difference across the eight areas in bacterial counts. Write a summary statement for your findings.

4. A research psychologist is investigating anorexia nervosa in adolescent females. He wishes to investigate the effects of psychodynamic counseling and behavioristic counseling (COUNSEL) along with three different levels

of diet (DIETTYPE: low calorie = 1, medium calorie = 2, and high calorie = 3) on weight gain (WT_GAIN). Thirty young women with anorexia nervosa are randomly assigned to these six different groups. The weight gain data are found in the file ANOREXIA. Are there main effects for either of the factors? Is there an interaction effect? If requested, answer these questions on the MYSTAT printout and write an appropriate summary statement for a journal article.

5. Using the KARACHI file, determine whether there is a significant difference between FOOD_TY$ and AREA with respect to bacterial counts. Is there an interaction between the two factors? Write a summary statement for this 2 x 8 ANOVA. You will have to make a new numerical variable for the text variable FOOD_TY$ to use it as a factor.

Notes

[1] There are several terms that have not been defined. In the title line of output, MYSTAT reports the name of the dependent variable and the number of subjects in the experiment. Also given are values for Multiple R and Squared multiple R. These two terms are defined in Chapter 9. The Sum-of-squares are used in calculating the ANOVA F statistic. Formulas for sum-of-squares and their interpretations are beyond the scope of this tutorial. If you divide the sum-of-squares by the degrees of freedom (DF) you will obtain the mean squares. Mean squares are variance estimates. The mean square for the factor GROUP estimates the variance in the dependent variable contributed by the treatment plus that contributed by error. The mean square for Error estimates the variance in the dependent variable produced by random error. If you divide the mean square for GROUP by the mean square for Error, you obtain the reported F value.

The mean square for GROUP (often called mean square between groups, MSB) estimates the variance in the dependent variable produced by both treatment effects and error. The mean square for Error (often called the mean square within groups, MSW) estimates the variance in the dependent variable produced by error. The formula for the F statistic is as follows:

$$F = \frac{MSB}{MSW} = \frac{\text{variance due to treatment effects} + \text{random error effects}}{\text{variance due to random error effects}}$$

If the differences in the independent variable (treatment) were not contributing to scatter or variance in the dependent variable, then the two mean squares would both measure the contribution of random error. Therefore, they should be equal, and if you divide them, their value should be 1. If the treatment doesn't produce differences in the dependent variable, then the expected value for the F statistic is 1. F values that are larger than 1 indicate that the independent variable is producing some variation in the dependent variable. If the F value is far enough away from 1, you reject the null hypothesis. The p value, as before, alerts you to whether the F statistic is far enough away from its expected value to be considered significant.

[2] The title line of output identifies the dependent variable and the number of subjects in the analysis. Under this title line is the Analysis of Variance source table. The row beginning with the word SOURCE lists the origin of the calculated values and the names of the calculations. The row beginning with the word DOSAGE reports the statistics associated with the first factor. In this row are the sum-of-squares, degrees of freedom (DF), mean square, F statistic, and probability due to the DOSAGE factor. You always compare the p value reported to the alpha level to determine if you reject the null hypothesis. Remember if the p value is less than the alpha level, reject H_0. The next line gives the same information for the second factor (therapy TYPE). The interaction (DOSAGE * TYPE) takes two lines because of the way the interaction effects are abbreviated in the SOURCE column. Finally the error effects are reported. Note that with an alpha of .05, the interaction is nonsignificant, so behavioral incidents are not shown to be jointly produced by dosage and therapy type. The type of therapy is not significant. Dosage is significant.

8

Graphing Data —Two or More Variables

In this chapter you will use the PLOT command to simultaneously graph two or more variables for each case in a data set. These graphs, called *scatterplots*, visually represent the relationship between two quantitative variables. To produce a scatterplot, the y, or vertical axis, is used to plot the dependent variable, and the x, or horizontal axis, is used to plot the independent variable. The point where these two values intersect for each case in the data set is marked with a dot (or another plot symbol). After you have produced several scatterplots, you will be able to make very good guesses about the direction and strength of these relationships.

Objectives

After finishing this tutorial you should be able to

■ Produce simple two-variable x-y scatterplots

■ Determine whether the data form a functional or statistical relationship

■ Set the axis scales

■ The PLOT Command

The **PLOT** command is used to produce scatterplots. The general form of the PLOT command is

> PLOT <VAR1> [, <VAR2>, <...>] * <VAR3> [/SYMBOL = <VAR$>
> '<CHAR>' YMAX = <#>, YMIN = <#>, XMAX = <#>, YMAX = <#>,
> LINES = <#>]

The PLOT command causes a scatterplot to be made with the variable listed before the asterisk graphed on the *y*-axis and the variable listed after the asterisk graphed on the *x*-axis. If more than one variable precedes the asterisk, a separate scatterplot is made for each of those variables. The SYMBOL option is used to replace the default dot used by MYSTAT with a symbol stored in a variable in the data (VAR$) or by a specific character (CHAR). If a character is used, it must be enclosed by single quotes. Because the PLOT command uses two variables, minimums and maximums can be set separately for each with the XMIN, XMAX, YMIN, and YMAX options. MYSTAT does a good job selecting the scales along the axes so that the screen is filled with points. You may want to set your own limits when you want to force scales to begin at zero or when you want to make the range of the scales the same for two or more plots. Finally, the LINE option determines the number of lines used to make the vertical axis.

Following are some valid PLOT commands:

- **PLOT READ * IQ/SYMBOL = RACE$**   produces a plot of reading scores on the *y*-axis and IQs on the *x*-axis. Each dot would be replaced with the first letter from the RACE$ variable.

- **PLOT Y1, Y2 * X/SYMBOL = '1', '2'**   produces two scatterplots on the values Y1 and Y2 graphed on the *y*-axis, and the value X graphed on the *x*-axis. Y1 values would appear as '1,' Y2 values as '2.'

- **PLOT Y * X/YMIN = 0, YMAX = 1, XMIN = 1.5, XMAX = 7.5**   plots the values Y versus X and sets scales on both axes.

- **PLOT X * Y/LINES = 75**   produces a plot with the vertical axis composed of 75 lines.

■ Two-Variable Scatterplots

Scatterplots are pictures of the relationship between variables. In this tutorial, you will produce a scatterplot showing a simple relationship between two continuous variables. In Chapter 2, we discussed the SCHOOL file. Assume you are a school psychologist interested in graphically displaying the relationship between the children's scores on the reading test (READ) and their full-scale intelligence quotients (FSIQ). Indeed, in psychology, intelligence quotients (IQs) are primarily used to predict academic achievement scores.

The following steps will produce the scatterplot:

1. From the main command screen, type **USE B:SCHOOL**, then press **[Enter]**.

2. Press **[Enter]** to return to the main command screen from the variable name screen.

   There are no mandates for determining which of the two variables in the PLOT command is called the dependent variable and which is called the independent variable. However, if you are going to use one variable to predict values for the other, the variable used to predict is referred to as the *predictor* variable and it becomes the independent variable in the study. The variable with estimated values is called the *criterion* variable and it becomes the dependent variable in the study. The achievement variable (READ) is the logical choice for the criterion variable because this is what FSIQ predicts. Traditionally, criterion variables are graphed along the *y*-axis, so we'll list READ first in the PLOT command.

   ☞ Remember, if you want to print analyses, including graphs, you must use the OUTPUT @ command before conducting the analyses.

3. Type **PLOT READ * FSIQ**, then press **[Enter]**.

   Your output should look like Figure 8.1.

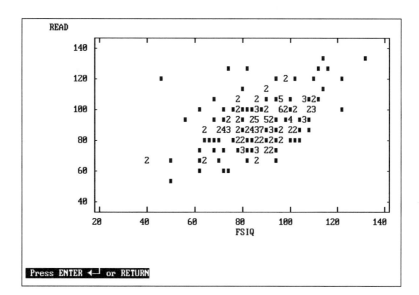

Figure 8.1
Scatterplot of FSIQ predicting READ

You have created a scatterplot for two variables. The boxes represent a single pair. If two dots are plotted on top of each other, it appears as the number 2; if three dots are plotted on top of one another, it appears as a 3; and so on.

■ Scatterplot Interpretation

Relationships between quantitative variables differ in direction and strength. These relationships can be visualized using scatterplots. If it is a *linear relationship,* that is, if a straight line can approximate this relationship, then the *Pearson product-moment correlation coefficient,* or simply the *correlation,* is a number used to describe the relationship. You can determine the direction of the correlation by looking at the general orientation of the scatterplot. If smaller values for the x variable are frequently paired with smaller values for the y variable and larger values for the x variable are paired with larger values for the y variable, the scatterplot will run from the lower-left corner of the graph to the upper-right corner. This indicates a *positive* relationship—as one variable's values become larger, the other variable's values also become larger. If smaller values for the x variable are frequently paired with larger y values and larger x values are paired with smaller y values, the scatterplot will run from the upper-left corner to the lower-right corner of the graph. This indicates a *negative* relationship. In Chapter 9 we will use MYSTAT to calculate the Pearson correlation coefficient between variables. If the relationship is positive, the correlation will be a positive number; if the relationship is negative, the correlation will be negative. Figures 8.2a-b illustrate the directions of scatterplots.

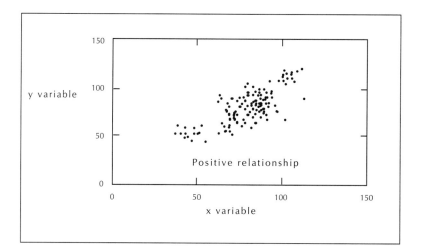

Figure 8.2a
Scatterplot of a
positive relationship

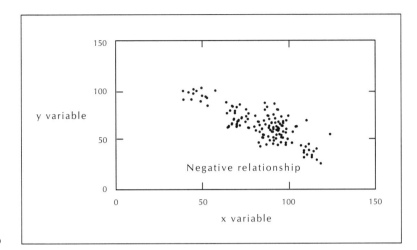

Figure 8.2b
Scatterplot of a
negative relationship

What is the direction of the relationship in Figure 8.1 between FSIQ and READ?
The second component of a relationship is its strength, which indicates the closeness of the relationship. Scatterplots are excellent tools for estimating the strength of a relationship. Pearson product-moment correlations measure the strength of linear relationships. If all the data points fit exactly on a straight line, then the relationship is perfect, and the correlation equals ±1 (depending on the direction). If the data points do not conform at all to a linear pattern, there is not a linear relationship and the Pearson correlation equals zero. Figures 8.3a-d depict different linear relationships.

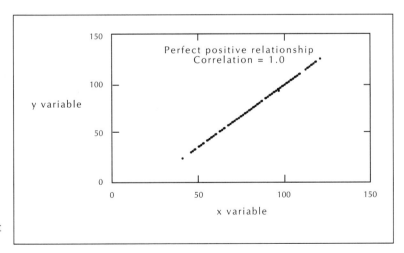

Figure 8.3a
Scatterplot of a perfect
positive relationship

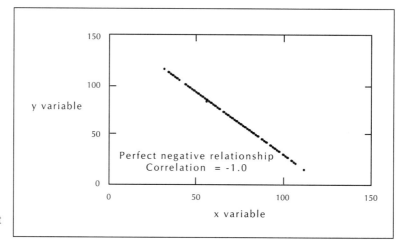

Figure 8.3b
Scatterplot of a perfect
negative relationship

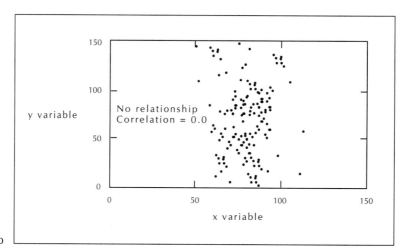

Figure 8.3c
Scatterplot in which
there is no relationship

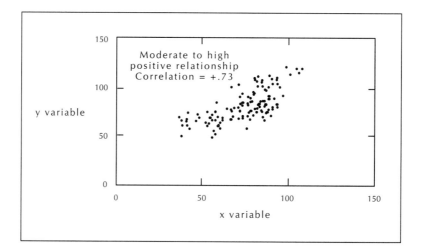

Figure 8.3d
Scatterplot of a
moderate positive
relationship

What would you guess the correlation to be for Figure 8.1? Would it be closer to .70 or closer to 0.0?

One advantage of using scatterplots is that you can sometimes see relationships the Pearson correlation underestimates. Pearson correlations underestimate relationships that aren't linear. Figure 8.4 illustrates data that are perfectly matched by a sine curve pattern. The Pearson correlation is also given in Figure 8.4. Your eyes can detect the strong relationship; the Pearson correlation cannot.

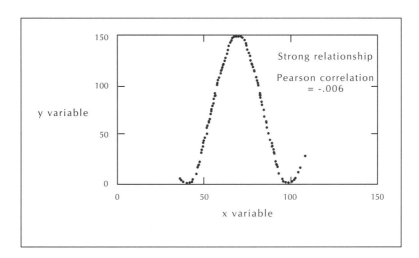

Figure 8.4
Scatterplot of a sine curve pattern

■ Functional and Statistical Relationships

When every value of one variable (x) determines a unique value of another (y), the variables have a functional relationship. For example, if y is the dependent variable and x is the independent variable, and y is determined by $y = 2x$, then a functional relationship exists. When graphs show functional relationships, the points fall directly on the function line or curve. Figure 8.5 is a graph of the functional relationship $y = \sqrt{x}$.

Statistical relationships are not perfect. When you look at the graph of a statistical relationship, the data points will not fall perfectly on a line or curve (see Figures 8.1, 8.2a, 8.2b, 8.3c, 8.3d).

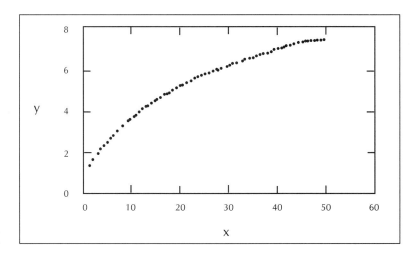

Figure 8.5
Scatterplot of $y = \sqrt{x}$.

■ Identifying Scatterplot Outliers

Another advantage of producing scatterplots is that data points far away from the majority are easily detected. These outside values are called *outliers*. Outliers are isolated points in the data swarm. In the scatterplot you produced in Figure 8.1, the point by itself in the upper left-hand quadrant of the graph is an outlier.

Outliers are often caused by errors in data entry. For example, the researcher might type 1400 instead of the correct value of 140. If outliers are not produced by a data entry error, they are often unique cases in the data set and are worthy of individual study to determine why they are so different.

———— Exercises

1. Produce a scatterplot from the SCHOOL file that uses VIQ to predict SPELL scores. Look at the two extreme outliers in the scatterplot on the SPELL variable. Can you determine which cases produced these scores?

2. If the SPELL scores must range from 45 to 145, what is the likely cause of these outliers?

3. Correct these scores to read 90 on SPELL, save the corrected data set as SCHLCORR, and produce a corrected scatterplot of VIQ predicting SPELL.

4. Delete all the cases from the SCHOOL data that have FSIQ > 110 and FSIQ < 85. Save this data set as DELETE. Construct a scatterplot in which FSIQ is on the *x*-axis and READ is on the *y*-axis. Does the shape of the scatterplot change from the first time these variables were graphed? Why or why not? (Hint: Use the IF…THEN LET command.)

5. Using the STATES file, produce a scatterplot between population density and summer temperature. Is there a linear relationship between these two variables?

6. When ocean navigators use sextants and the sun to find their position, they must correct their sextant readings using a Dip correction. These Dip values change depending on the height of the observer's eye above the water. Thus, a navigator on a small ship close to the water's surface would use a different correction value than a person standing on the deck of a large freighter. A table of Dip corrections is provided in the file DIPS. Plot the correction (DIP) on the *y*-axis against the height of the observer's eye (HEIGHT) on the *x*-axis. Is there a relationship between these values? Is this relationship best described by a line or a curve? Is this a functional or statistical relationship?

7. Rank the DIPS data using only the DIP variable in the ranking. Title this new file DIPSRANK. Plot the ranked DIP variable on the *y*-axis against HEIGHT on the *x*-axis. Is the relationship changed by the rank ordering of this single variable? If so, describe the change.

8. Rank the DIPSRANK data using HEIGHT as the ranking variable. Title this new file DIPSRNK2. Now both DIP and HEIGHT are in rank order. Plot DIP on the *y*-axis and HEIGHT on the *x*-axis. Does the relationship change when both variables are in rank order? If so, describe the change.

Correlation and Regression

As noted in Chapter 8, a correlation coefficient measures the strength of association between variables. A value of zero indicates that there is no association. Values of either -1 or +1 indicate perfect associations between the variables. The MODEL and ESTIMATE commands are combined to compute *bivariate* (one predictor and one criterion) and *multiple regression* (more than one predictor and one criterion) solutions. The PEARSON command computes correlations between many pairs of variables and other correlations like the Spearman rank correlation. The MODEL and ESTIMATE commands compute estimates of regression coefficients and test their significance. They are used in combination to effect a regression analysis.

Objectives

At the end of this chapter you should be able to

- Determine whether to calculate a Pearson correlation or a Spearman correlation or to conduct a regression analysis
- Calculate Pearson correlations between any two numeric variables
- Calculate Spearman correlations and understand the difference between the Pearson and Spearman correlations
- Compute and interpret bivariate regression analyses
- Compute and interpret multiple regression analyses
- Detect outliers in regression analysis and understand Cook's D statistic, leverage estimates, and studentized residuals
- Use common plots to analyze residuals

■ Determining Whether to Use the PEARSON Command or a Regression Analysis

Correlations are for measuring association, regression is for prediction. As we said in Chapter 8, the Pearson product-moment correlation coefficient is a number that measures the linear relationship between quantitative variables. The PEARSON command computes Pearson correlations and other correlations that use the same formula. The other correlations have various names, but because they all use the same formula they are considered to be in the "Pearson family."

The *Spearman rank correlation* is a number that measures the relationship between rank-ordered variables. It is simply a Pearson correlation on data whose quantitative values are replaced by their rank orders.

The other Pearson family coefficients all involve one or more dichotomous variables. *Dichotomous variables*, also called binary variables, have only two values. The **PEARSON** command produces correlations between dichotomous variables, but it does not test them for statistical significance. Therefore, PEARSON command results are generally used for descriptive purposes.

The **MODEL** and **ESTIMATE** commands enable you to conduct a regression analysis. This means calculating the necessary inferential statistics so that you use the value of the predictor variable to estimate the criterion value. If you use a single predictor to estimate a single criterion, you are conducting a *bivariate regression*. If you use several predictors to estimate a single criterion, you are executing a *multiple regression*.

Figure 9.1 illustrates the decision model for determining whether to use the PEARSON command to calculate a Pearson or Spearman correlation or the regression analysis procedures to conduct bivariate or multiple regressions.

■ Correlation Coefficients

Many correlation coefficients are available to researchers. The two most popular are the Pearson product-moment correlation coefficient (named after the statistician Karl Pearson) and the Spearman rank coefficient (named after the British psychologist Charles Spearman). Traditionally, the Pearson correlation has been used to indicate the degree of association between two quantitative variables, while the Spearman correlation has been used to indicate the degree of association between two variables that are ranks. As you will discover, these two correlations may actually use the same formula. Several other correlations also employ the Pearson formula. Two that you should know are the *point biserial* correlation (r_{pb}) in which one of the variables is measured as a dichotomy and the other is quantitative, and the *phi* (ϕ) coefficient in which both variables are measured as dichotomies.

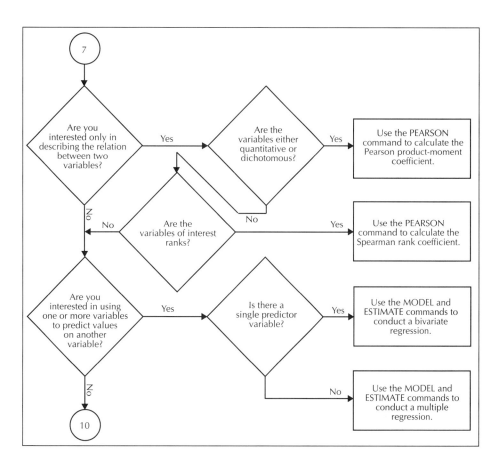

Figure 9.1
Decision model for
PEARSON and
regression analysis

Pearson Product-Moment Correlation

The Pearson product-moment correlation is symbolized in most textbooks with an *r* and is appropriate when both variables are quantitative. If one or more variables are dichotomous, the same steps and the same formula are used to calculate the correlation. The only difference is that the name of the coefficient changes.

Calculation of *r*

To measure the degree to which several variables in the SCHOOL file are related, you will use the PEARSON command to calculate the Pearson correlations between each of them. The general form of the PEARSON command is:

PEARSON [<VAR1>, <VAR2>, <...>/PAIRWISE I LISTWISE]

If your data set has missing values you can instruct MYSTAT to ignore the case using the PAIRWISE or LISTWISE deletion option. These options tell MYSTAT to delete the case from the calculation. The default is to delete cases using the LISTWISE option. Therefore, if no option or the LISTWISE option is chosen, every case that has a missing value on any of the selected variables will be dropped from the calculation of every correlation. Using this option, every correlation is calculated using the same number of subjects. The PAIRWISE option drops a case from a calculation only when the value is missing for one of the variables in the pair. If you choose the PAIRWISE option and you have missing values, the output is divided into two parts. The first contains the correlation coefficients and the second contains whole numbers indicating the number of cases used in the calculation of each coefficient.

To produce this Pearson correlation matrix,

1. From the main command screen, type **USE B:SCHOOL**, then press **[Enter]** twice to load the data into MYSTAT.

2. Type **PEARSON VIQ, PIQ, FSIQ, READ, SPELL, ARITH, AGGRESS, WITHDRAW**.

   These are the variables that will be used by the PEARSON program for calculating the coefficients. If you do not choose any variables, MYSTAT calculates correlations between every numeric variable. The PEARSON command calculates a correlation for each possible pair of variables selected. The correlation for any variable with itself is always +1.0. If all the possible correlations were calculated regardless of the order of the variables, these values could be placed in a square table or matrix. Figure 9.2 illustrates this square matrix. Note that 1.0 is placed in each cell in which a variable is correlated with itself (on the diagonal).

	VIQ	PIQ	FSIQ	READ	SPELL	ARITH	AGGRESS	WITHDRAW
VIQ	1.0	.65	.91	.54	.49	.61	-.01	.26
PIQ	.65	1.0	.94	.34	.28	.52	-.03	.17
FSIQ	.91	.94	1.0	.49	.43	.62	-.01	.24
READ	.54	.34	.49	1.0	.83	.63	.14	.10
SPELL	.49	.28	.43	.83	1.0	.65	.15	.07
ARITH	.61	.52	.62	.63	.65	1.0	.03	.10
AGGRESS	-.01	-.03	-.01	.14	.15	.03	1.0	-.19
WITHDRAW	.26	.17	.24	1.0	.07	.10	-.19	1.0

Figure 9.2
A square correlation matrix

The correlation between any pair of variables is the same no matter what the order. For example, the correlation between VIQ and PIQ (.65) is identical to that between PIQ and VIQ. Because the correlation does not change when the order of the variables changes, this square matrix can be divided into two triangular matrices. The upper-right triangle of this matrix (shaded in the figure) provides the same information as the lower-left triangular matrix. Each column in the lower-left triangle contains values identical to those of the corresponding row in the upper-right triangle. Therefore, both triangles are not required. MYSTAT only prints the lower-left triangle and the diagonal values of 1.0.

3. Press [**Enter**].

Your output should look like Table 9.1.

PEARSON CORRELATION MATRIX

	VIQ	PIQ	FSIQ	READ	SPELL
VIQ	1.000				
PIQ	0.656	1.000			
FSIQ	0.914	0.904	1.000		
READ	0.542	0.344	0.493	1.000	
SPELL	0.176	0.095	0.156	0.181	1.000
ARITH	0.610	0.522	0.624	0.627	0.029
AGGRESS	-0.009	-0.029	-0.014	0.138	-0.025
WITHDRAW	0.261	0.170	0.241	0.099	-0.045

	ARITH	AGGRESS	WITHDRAW
ARITH	1.000		
AGGRESS	0.025	1.000	
WITHDRAW	0.099	-0.189	1.000

NUMBER OF OBSERVATIONS: 200

Table 9.1
PEARSON output in lower-left triangular form

☞ The PEARSON output is too wide to fit in a tidy triangle on the screen. The triangle is split into two sections and the second section is printed below the first.

Correlations between different variables are positioned in the off-diagonal locations. For example, the correlation between FSIQ and READ is 0.493. You might want to refer back to the scatterplot created in Figure 8.1 to see what a correlation of 0.49 looks like.

In Chapter 8, it was stated that the Pearson coefficient measures the degree of linear association between variables. Let's do a quick demonstration of this property.

Do the following steps:

1. From the main command screen, type **USE B:POWERS** and press [**Enter**].

   POWERS contains fifty cases with six named variables: X, Y1, Y2, Y3, Y4, Y5. The X variable is simply the case number (i.e., consecutive numbers from 1 through 50). Y1 is the X variable to the first power, which means that X and Y1 are identical. Y2 is the square of the X variable, Y3 is the cube of the X variable, and Y4 and Y5 are the fourth and fifth powers of the X variable. If you produce separate plots of each of these Y variables with respect to the X variable (place the Y variables on the y-axis), you will note that the degree of curvature for the plots increases. For example, Y1 versus X is a straight line, Y2 versus X is slightly curved, and Y5 versus X is highly curved.

   ☞ Remember, if you want to see the data, use the LIST command.

2. Press [**Enter**] to return to the main command screen.

   Now produce all the correlations between these variables.

3. Type **PEARSON** and press [**Enter**] to have correlations calculated on all the variables (see Table 9.2).

PEARSON CORRELATION MATRIX

	X	Y1	Y2	Y3	Y4
X	1.000				
Y1	1.000	1.000			
Y2	0.969	0.969	1.000		
Y3	0.919	0.919	0.986	1.000	
Y4	0.869	0.869	0.959	0.992	1.000
Y5	0.824	0.824	0.928	0.975	0.995

	Y5
Y5	1.000

NUMBER OF OBSERVATIONS: 50

Table 9.2
Pearson correlations for the POWERS data

All these data points are derived directly from one another and form functional relationships. This means that even though these relationships are curved, they are perfect. However, the only Pearson correlations that indicate perfect relationships are between variables that form linear scatterplots. Pearson correlations are higher between variables with closer exponents (these pairs form more linear scatterplots) and lower between variables whose exponents are further apart (these pairs form curved scatterplots). As you will see in the next section, if you want to measure associations that do not depend on an assumption of linearity, the Spearman coefficient is used.

Spearman Rank

The Spearman rank correlation is symbolized by r_s in most textbooks and it uses only rank order information. If a variable consists of the values 1, 3, 14, and 22 and was then ranked, the values would be 1, 2, 3, and 4. The Spearman correlation is simply a Pearson between two variables whose values are ranked. Indeed, it can be shown that the traditional calculation formula for the Spearman correlation is equivalent to the Pearson formula when one uses two sets of consecutive untied ranks. Tied values can be replaced by averaged ranks, which is how MYSTAT computes ranks with the RANK command. Using the Pearson formula on ranks averaged for ties produces a Spearman correlation coefficient that has been corrected for ties.

Pearson correlations can be 1 (or -1) only if the x and y values fall on a straight line in a scatterplot. Spearman correlations can be 1 only if the ranks of the x and y values are identical, and -1 only if the ranks of x are the reverse of the ranks of y. That is, the ranks would fall on a straight line in a scatterplot of ranks.

Calculating r_s

To calculate the Spearman coefficient, rank the variable values using the RANK command and then calculate a Pearson correlation using the PEARSON command on the ranked data set.

To conduct this analysis with the POWERS data,

1. Type **USE B:POWERS** and press **[Enter]** twice.
2. Type **SAVE B:POWERANK**, then press **[Enter]** to create a file in which the ranked data will be saved.
3. Type **RANK X, Y1, Y2, Y3, Y4, Y5**, then press **[Enter]** to rank the named variables and automatically save the ranked data to POWERANK. Press **[Enter]** again to return to the main command screen.
4. Type **USE B:POWERANK** and press **[Enter]** twice to load the ranked data.

   WARNING: *If you don't use this new data set you will be computing the same Pearson coefficients found in Table 9.2.*

5. Type **PEARSON**, then press **[Enter]**.

   Because all the data are in rank order, the correlations calculated are Spearman correlations even though MYSTAT labels the output as a Pearson correlation matrix (see Table 9.3). Because each of these variables formed a functional monotonic relationship with all the other variables, you should not be surprised to see the results produced in Table 9.3. While the POWERS variables all formed functional relationships (see Chapter 8), the Spearman correlation measures in addition the degree of association in monotonic statistical relationships.

PEARSON CORRELATION MATRIX

	X	Y1	Y2	Y3	Y4
X	1.000				
Y1	1.000	1.000			
Y2	1.000	1.000	1.000		
Y3	1.000	1.000	1.000	1.000	
Y4	1.000	1.000	1.000	1.000	1.000
Y5	1.000	1.000	1.000	1.000	1.000
	Y5				
Y5	1.000				

Table 9.3
Spearman correlations
for the POWERS data   NUMBER OF OBSERVATIONS: 50

This example points out a major similarity (use of the same formula) and a major difference (a linear versus a rank interpretation) between the two correlations.

■ Bivariate Regression Analysis

When you expect that two variables may be related, you might hypothesize that one of the variable's values can be used to predict the values of the second variable. This statistical procedure for prediction is called *regression analysis*, and the equation for the *best-fitting* straight line is called the *linear regression equation*. A *regression line* is the best straight line for predicting the *y* values from the *x* values in a scatterplot. If you draw your own line through the data swarm, then compute the vertical distance between the line and the dot for every case, square that distance, and then add those squared values together, you will get a number (sum of squared residuals) that indicates how closely your line fits the data. For all possible lines through the swarm, the best predicting line has the smallest sum of squared residuals. Figure 9.3 illustrates two cases to be squared and summed to get the value for this line. No other straight line would fit better (that is, give a smaller value for these squares) than the regression line. The Pearson correlation measures how closely the data values match this straight regression line.

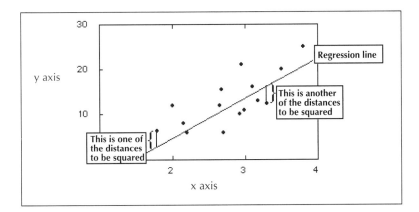

Figure 9.3
Scatterplot with the regression line

A simple example will illustrate the regression technique. Earlier you saw that the Pearson correlation between VIQ and READ scores in the SCHOOL file was .542. You might construct the best straight line that fits the data to make predictions, and then conduct an inferential test to determine if this straight line is a significant predictor of the criterion (READ). All of this can easily be done with the MODEL and ESTIMATE commands.

The MODEL Command

The MODEL command calculates regression lines. When used in conjunction with the ESTIMATE command, it conducts a regression analysis. The general form of the MODEL command is

> **MODEL <VAR1> [, <VAR2>, <...>] = [CONSTANT +] <VAR3>
> [+ <...> + <VAR4> * <VAR5>]**

The optional CONSTANT term is the place where the regression line crosses the *y*-axis when the predictor value equals zero. The following are all valid MODEL commands:

- **MODEL Y = X** calculates a regression without the constant.
- **MODEL Y = CONSTANT + X** conducts a simple bivariate regression.
- **MODEL Y = CONSTANT + X + Z** conducts a multiple regression.
- **MODEL Y = CONSTANT + X + Z + X * Z** conducts a multiple regression with an interaction term for the variables X and Z included.
- **MODEL Y = CONSTANT + X + X * X** conducts a polynomial regression.

☞   MODEL commands may also be used in analysis of variance problems. These are discussed in a note at the end of this chapter.[1]

The Problem

You are a public school teacher and you wish to find out whether your students' verbal intelligence scores (VIQ) can predict their reading scores (READ). If the prediction is statistically significant, then you wish to construct the best equation for taking those intelligence scores and predicting how well other children can be expected to do on the same reading test.

The Solution

1. Write the null and alternative hypotheses.

In regression analysis, you are asking whether the increase in accuracy of prediction of the individual data values when using the regression equation is significantly better than predicting the data values without the equation. We will state the null and alternative hypotheses using words.

H_0: The regression equation does not significantly improve our predictions.

H_1: The regression equation significantly improves the predictions.

2. Set the alpha level.

Set your alpha level to .05 for this problem.

3. Collect the data.

The data are contained in the SCHOOL file.

   1. Type **USE B:SCHOOL**, then press **[Enter]** twice to load the data into MYSTAT.

4. Calculate the statistic.

   1. Type **MODEL READ = CONSTANT + VIQ**, then press **[Enter]** to calculate the regression.
   2. Type **ESTIMATE** and press **[Enter]** to complete the regression analysis.

   You will obtain results shown in Table 9.4.[2]

Look at the "Analysis of Variance" summary table at the bottom of the output. Here MYSTAT is conducting a significance test using the F statistic to determine whether the regression equation significantly improves your predictions. Do you remember your null hypothesis from step 1? All you need to consider is the reported p value (labeled P in the output). If the p value is less than the alpha level set in step 2, reject H_0.

5. Decide whether or not to reject the null hypothesis.

Because the p value reported is less than .05, reject H_0.

DEP VAR: READ   N: 200   MULTIPLE R: .542   SQUARED MULTIPLE R: .294
ADJUSTED SQUARED MULTIPLE R: .290   STANDARD ERROR OF ESTIMATE: 12.318

VARIABLE	COEFFICIENT	STD ERROR	STD COEF	TOLERANCE	T	P (2 TAIL)
CONSTANT	42.262	5.086	0.000	.	8.309	0.000
VIQ	0.530	0.058	0.542	.100E+01	9.077	0.000

ANALYSIS OF VARIANCE

SOURCE	SUM-OF-SQUARES	DF	MEAN-SQUARE	F-RATIO	P
REGRESSION	12502.171	1	12502.171	82.401	0.000
RESIDUAL	30041.329	198	151.724		

Table 9.4
SCHOOL regression results of VIQ predicting READ

6. Write a summary statement.

In this case the researcher might report: "For these children, verbal IQ was found to be a significant predictor of reading scores (r = .542, F = 82.4, df 1, 198, p < .0005)." The generated regression equation is Expected Reading = 42.262 + 0.53VIQ. To place confidence intervals around the expected reading score the standard error of estimate of 12.318 can be used.

☞ In the summary statement the regression equation is in the form of $\hat{Y} = a + bX$ where $\hat{Y}$ is a predicted value, a is a constant, X is the predictor variable, and b is the regression coefficient. This is how regression equations are typically reported.

Assumptions for Regression Analysis

Three requirements must be met before using ordinary linear regression: the data points must be uniformly scattered around the regression line (called the *homoscedasticity of variance*), the data values must be independent of each other, and the true relation between Y and X should be linear. In addition, in order for the tests of significance to be appropriate, the residuals must come from a normal distribution. These assumptions are most often verified by looking at a scatterplot.

■ Multiple Regression Analysis

When several quantitative independent variables are used to estimate a single criterion score, the procedure is called *multiple regression analysis*. Instead of calculating a regression line that fits the data, you calculate a regression plane. Figure 9.4 shows a regression plane in which verbal intelligence scores (VIQ) are combined with performance IQs (PIQ) to predict the dependent variable, reading achievement scores (READ). The regression surface is no longer a line but a plane. More than two predictors cannot be easily graphed because the graph would contain more than three dimensions. MYSTAT can conduct a multiple regression using any number of independent variables, but MYSTAT will not produce three-dimensional graphs. Figure 9.4 was produced using SYSTAT, a commercial statistics package.

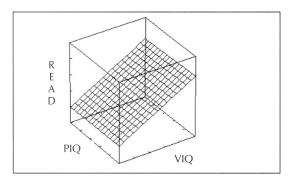

Figure 9.4
A regression plane
created using SYSTAT

The Problem

Assume you are interested in discovering whether VIQ and PIQ scores combine to predict READ in children. You can use the SCHOOL data set to answer this question.

The Solution

Using the six-step solution, conduct the multiple regression analysis.

1. Write the null and alternative hypotheses.

H_0: The regression equation using both VIQ and PIQ does not significantly improve our predictions.

H_1: The regression equation using both VIQ and PIQ significantly improves the predictions.

2. Set the alpha level.

Set the alpha level at .05 for this problem.

3. Collect the data.

    1. Type **USE B:SCHOOL** and press **[Enter]** twice to load the data into MYSTAT.

4. Calculate the statistic.

    1. Type **MODEL READ = CONSTANT + VIQ + PIQ** and press **[Enter]** to conduct a regression on VIQ and PIQ.

    2. Type **ESTIMATE** and press **[Enter]** to calculate the regression. Your output should look like Table 9.5.

DEP VAR: READ   N: 200   MULTIPLE R: .542   SQUARED MULTIPLE R: .294
ADJUSTED SQUARED MULTIPLE R: .287   STANDARD ERROR OF ESTIMATE: 12.347

VARIABLE	COEFFICIENT	STD ERROR	STD COEF	TOLERANCE	T	P (2 TAIL)
CONSTANT	42.882	5.627	0.000	.	7.620	0.000
VIQ	0.543	0.078	0.556	0.5697305	7.006	0.000
PIQ	-0.019	0.075	-0.021	0.5697305	-0.260	0.795

ANALYSIS OF VARIANCE

SOURCE	SUM-OF-SQUARES	DF	MEAN-SQUARE	F-RATIO	P
REGRESSION	12512.502	2	6256.251	41.040	0.000
RESIDUAL	30030.998	197	152.442		

Table 9.5
Multiple regression output for VIQ and PIQ predicting READ

5. Decide whether or not to reject the null hypothesis.

Because the p value for the F reported in the ANOVA table is less than .05, reject H_0.

6. Write a summary statement.

In this case the researcher might report: "For these children, Verbal IQ and Performance IQ scores were found to significantly predict reading scores (R = .542, F = 41.04, df = 2, 197, p < .0005)." The generated regression equation is Expected Reading = 42.882 + 0.543VIQ - 0.019PIQ. To place confidence intervals around the expected reading score, the standard error of estimate (12.347) can be used.

There is a complication. The overall equation is significant, but the PIQ coefficient is not. We cannot conclude that both variables are needed to predict READ. Look back at Table 9.4. Note that the multiple correlation (.542) is the same for both models indicating that the prediction is not significantly improved by adding PIQ. At the same time, we cannot conclude that PIQ is not a good predictor of READ. Try regressing READ on PIQ instead of VIQ. The multiple correlation is .344, somewhat less than before but still significant.

Note that the correlation between VIQ and PIQ is .656. Since each variable is related to READ separately, this means that either variable is doing part of the work of the other when predicting READ. When both are put together, the stronger predictor (VIQ) is significant and the other predictor (PIQ) has no useful prediction to add independently. When this condition exists, it is difficult to make simple statements about the predictors because the relationship between them masks their separate ability to predict.

The t-test values in the upper portion of the output can be used to test whether each individual predictor is contributing to the prediction. Again, you look at these p values to make the decision. VIQ is a significant predictor, but PIQ with a p value of .795 is not. Look at how little PIQ changed the output. It would be important for parents, teachers, and psychologists to know how little a child's Performance IQ adds to VIQs when predicting some measures of academic achievement. As discussed previously, the tolerance value indicates how unique each variable in the regression is. If independent variables are not unique, they will be correlated with one another. The square of their correlation indicates how much variance they share. Look back at Table 9.1. The correlation between VIQ and PIQ is .656. If you square this value, you get .430336. This means that 43% of the variance in the predictors are shared. The unique variance of each independent variable is 1 - .430336 = .569664. How does this value agree with the tolerance figure reported in Table 9.5? The difference is due to rounding error and the number of decimals displayed in Table 9.1.

■ Outlier Detection

When you conduct bivariate regressions, it is a relatively simple matter to use box plots, stem-and-leaf plots, and scatterplots to detect outlying cases. With two or more predictor variables, graphic outlier detection becomes more and more problematic. With multiple regressions you need to interpret indices that detect outliers. Outliers in data sets can be produced in three ways: the values for the predictors or independent values may be unusual; the criterion or *y* values may be unusual; or a single data point may have an undue impact on the regression equation.

Leverage

A *leverage* measure indicates whether the predictor values for a specific case are aberrant. The leverage value measures the distance between predictor values for any specific case and the average of the predictor values for all the cases in the study. If the leverage value is large, it indicates that the predictor values for this case are far from the mean of all the predictors.

There are several rules for using leverage values to identify outliers. (1) A leverage is often considered large if it is more than twice as large as the average leverage value. (2) Consider a leverage value greater than .5 to be large. (3) Finally, some researchers look for gaps between where most of the leverages are concentrated and a few stray leverages. The stray leverages are considered large.

Studentized Deleted Residuals

Studentized deleted residuals indicate how different predicted values are from their corresponding actual values. When the regression analysis is completed, a regression equation of the form $\hat{Y} = a + b_1X_1 + b_2X_2 + \ldots$ is calculated. You can calculate expected criterion values ($\hat{Y}$) for any set of predictor values by substituting the predictor scores into the regression equation. For each case both the actual criterion score (Y) and the predicted criterion score ($\hat{Y}$) are available. Using these two values, you can calculate the residual (Y - $\hat{Y}$). The residual is the amount by which the regression equation missed the actual criterion value. (See Figure 9.3 and the earlier discussion about errors and residuals.)

It is difficult to interpret the absolute value of these residuals to detect outliers for two reasons. The first is a problem of scale. If the original variables are measured with numbers whose scale is small (e.g., the numbers might vary from 1 through 10 with a mean of 5), a residual of 15 might be quite large. Fifteen in this example is three times the mean. However, if the variables are measured on a scale of 1 to 1,000 with a mean of 500, missing the criterion by 15 points might indicate a very close approximation. The second problem with using absolute values of residuals is that if the criterion value is abnormal, it unduly influences the regression equation so that the regression line is pulled closer to the criterion value than it would be if this case were eliminated from the analysis. These interpretation problems are solved by using *studentized deleted residuals*, which are transformed residual scores. The transformation standardizes residuals and calculates the value without including particular cases in the analysis. Thus, studentized deleted residuals take care of both interpretation difficulties.

Studentized deleted residuals take the form of a *t* distribution. Judd and McClelland (1989) have advice for interpreting studentized deleted residuals:

> We can suggest an easy rule of thumb for identifying outliers. For reasonably large *n*, approximately 95% of the studentized deleted residuals (which have a *t* distribution) will be between -2 and +2. Thus, studentized deleted residuals with an absolute value less than 2 are not surprising, and we will not consider them to be unusual. However, if the absolute value of the studentized deleted residual is greater than 2, then it probably deserves another look because values that large should occur less than 5% of the time (approximately). Only about 1% of the studentized deleted residuals should be less than -3 or greater than +3. So, if the absolute value of the studentized deleted residual is greater than 3 careful attention to that observation is required. Finally, absolute values of studentized deleted residuals greater than 4 ought to be extremely rare; if the absolute value is greater than 4, then all alarm bells ought to sound (p. 225).

Cook's D

Cook's D measures the influence a case has on the regression. If you eliminate one case from a regression analysis and recompute the results, the value of the regression coefficients should not change dramatically. If that case has undue influence on the regression equation, then removing that case dramatically changes the values of the computed regression coefficients. Cook's D measures this influence.

Again, we use several rules to determine when Cook's D is too large. One rule is to consider any value greater than 1 or 2 as atypical. Another rule is to look for gaps between the small Cook's D values and the larger ones. Finally, according to Judd and McClelland (1989), a frequent rule is to consider Cook's D values large if they are greater than 4/(Number of predictor variables + 1) * (number of subjects). If the SCHOOL data is used with VIQ predicting READ, then values above 4/(2)*(200) or .01 would be considered large.

MYSTAT flags unusual cases, using approximate 80% confidence intervals on Cook's D leverage and studentized residuals if the SAVE command is used in the regression analysis procedure. This may identify a large number of cases as possible outliers. It is better for the researcher to look at the cases flagged as possible outliers and make a decision then for MYSTAT to fail to identify one. Neter, Wasserman, and Kutner (1990) and Judd and McClelland (1989) provide excellent discussions about outlier detection.

If you include the SAVE command in the regression procedure, MYSTAT creates a file containing the variables you used in your equation, the estimated criterion values (ESTIMATE), residuals (RESIDUAL), the standard error of prediction (SEPRED), leverage (LEVERAGE), Cook's D (COOK), and externally studentized deleted residuals (STUDENT).

In addition to saving the file, when the SAVE command is used with a linear model, the output includes a listing of cases with extreme studentized residuals (or leverage values), which are potential outliers. It also prints the Durbin-Watson statistic and the first-order autocorrelation coefficient. The *autocorrelation coefficient* measures the correlation of the errors or residuals over time. The *Durbin-Watson statistic* tests whether the autocorrelation equals zero. These measures are important in business and economics where regression analysis is used to solve problems involving time series data. This topic is not discussed in this tutorial; however, an excellent discussion of autocorrelation and the Durbin-Watson statistic can be found in Neter, Wasserman, and Kutner (1990).

Rerun the regression in which you predict READ scores given VIQ and include a SAVE command.

1. Type **USE B:SCHOOL** and press **[Enter]** twice.

   ☞   This step is not necessary if the SCHOOL file is currently loaded.

2. Type **SAVE B:SCHLRSID** and press **[Enter]** to create a file in which the studentized deleted residuals will be stored.

   ☞   The SAVE command may be placed either before or after the MODEL command.

3. Type **MODEL READ = CONSTANT + VIQ** and press **[Enter]**.

4. Type **ESTIMATE** and press **[Enter]**.

   The output from this analysis (see Table 9.6) details which cases are possible outliers and why MYSTAT believes they may be outliers.

DEP VAR: READ   N: 200   MULTIPLE R: .542   SQUARED MULTIPLE R: .294
ADJUSTED SQUARED MULTIPLE R: .290   STANDARD ERROR OF ESTIMATE: 12.318

VARIABLE	COEFFICIENT	STD ERROR	STD COEF	TOLERANCE	T	P (2 TAIL)
CONSTANT	42.262	5.086	0.000	.	8.309	0.000
VIQ	0.530	0.058	0.542	.100E+01	9.077	0.000

ANALYSIS OF VARIANCE

SOURCE	SUM-OF-SQUARES	DF	MEAN-SQUARE	F-RATIO	P
REGRESSION	12502.171	1	12502.171	82.401	0.000
RESIDUAL	30041.329	198	151.724		

```
WARNING: CASE     4 IS AN OUTLIER     (STUDENTIZED RESIDUAL =     -1.452)
WARNING: CASE     5 HAS LARGE LEVERAGE          (LEVERAGE =       .042)
WARNING: CASE     8 IS AN OUTLIER     (STUDENTIZED RESIDUAL =      2.048)
WARNING: CASE     9 IS AN OUTLIER     (STUDENTIZED RESIDUAL =     -1.626)
WARNING: CASE    11 IS AN OUTLIER     (STUDENTIZED RESIDUAL =      2.225)
WARNING: CASE    13 IS AN OUTLIER     (STUDENTIZED RESIDUAL =      2.066)
WARNING: CASE    23 IS AN OUTLIER     (STUDENTIZED RESIDUAL =      1.419)
WARNING: CASE    24 IS AN OUTLIER     (STUDENTIZED RESIDUAL =     -1.658)
WARNING: CASE    29 IS AN OUTLIER     (STUDENTIZED RESIDUAL =      1.335)
WARNING: CASE    32 HAS LARGE LEVERAGE          (LEVERAGE =       .042)
WARNING: CASE    42 IS AN OUTLIER     (STUDENTIZED RESIDUAL =      2.264)
WARNING: CASE    44 IS AN OUTLIER     (STUDENTIZED RESIDUAL =     -1.825)
WARNING: CASE    50 IS AN OUTLIER     (STUDENTIZED RESIDUAL =     -1.428)
WARNING: CASE    52 IS AN OUTLIER     (STUDENTIZED RESIDUAL =     -1.621)
WARNING: CASE    57 IS AN OUTLIER     (STUDENTIZED RESIDUAL =      1.405)
WARNING: CASE    60 IS AN OUTLIER     (STUDENTIZED RESIDUAL =     -1.384)
WARNING: CASE    63 HAS LARGE LEVERAGE          (LEVERAGE =       .039)
WARNING: CASE    63 IS AN OUTLIER     (STUDENTIZED RESIDUAL =      1.876)
WARNING: CASE    67 IS AN OUTLIER     (STUDENTIZED RESIDUAL =     -1.538)
WARNING: CASE    69 IS AN OUTLIER     (STUDENTIZED RESIDUAL =     -1.772)
WARNING: CASE    71 IS AN OUTLIER     (STUDENTIZED RESIDUAL =     -1.426)
WARNING: CASE    93 IS AN OUTLIER     (STUDENTIZED RESIDUAL =     -1.592)
WARNING: CASE    98 IS AN OUTLIER     (STUDENTIZED RESIDUAL =      1.721)
WARNING: CASE   102 IS AN OUTLIER     (STUDENTIZED RESIDUAL =      1.805)
WARNING: CASE   104 IS AN OUTLIER     (STUDENTIZED RESIDUAL =      1.650)
WARNING: CASE   121 IS AN OUTLIER     (STUDENTIZED RESIDUAL =      1.401)
WARNING: CASE   125 IS AN OUTLIER     (STUDENTIZED RESIDUAL =     -1.576)
WARNING: CASE   138 IS AN OUTLIER     (STUDENTIZED RESIDUAL =     -1.334)
WARNING: CASE   154 IS AN OUTLIER     (STUDENTIZED RESIDUAL =     -1.467)
WARNING: CASE   169 HAS LARGE LEVERAGE          (LEVERAGE =       .034)
WARNING: CASE   174 IS AN OUTLIER     (STUDENTIZED RESIDUAL =     -1.603)
WARNING: CASE   181 IS AN OUTLIER     (STUDENTIZED RESIDUAL =     -1.694)
WARNING: CASE   182 HAS LARGE LEVERAGE          (LEVERAGE =       .041)
```

Table 9.6 cont.

WARNING: CASE	182 IS AN OUTLIER	(STUDENTIZED RESIDUAL =	4.456)
WARNING: CASE	185 IS AN OUTLIER	(STUDENTIZED RESIDUAL =	2.790)
WARNING: CASE	186 IS AN OUTLIER	(STUDENTIZED RESIDUAL =	-1.811)
WARNING: CASE	192 IS AN OUTLIER	(STUDENTIZED RESIDUAL =	2.160)
WARNING: CASE	195 IS AN OUTLIER	(STUDENTIZED RESIDUAL =	1.433)
WARNING: CASE	196 IS AN OUTLIER	(STUDENTIZED RESIDUAL =	1.971)
WARNING: CASE	198 IS AN OUTLIER	(STUDENTIZED RESIDUAL =	1.445)
WARNING: CASE	199 IS AN OUTLIER	(STUDENTIZED RESIDUAL =	3.891)

Table 9.6
Regression results of VIQ predicting READ with the SAVE command incorporated

DURBIN-WATSON D STATISTIC   1.961
FIRST ORDER AUTOCORRELATION   .018

You may want to investigate the SCHLRSID file, which was created to look at the variables saved from this analysis. Figure 9.5 contains the first several variables for the initial ten cases.

```
MYSTAT Editor
 Case   ESTIMATE    RESIDUAL    LEVERAGE      COOK      STUDENT
   1      87.297      -3.297        .005       .000       -.268
   2      94.185      -1.185        .008       .000       -.096
   3      86.767     -14.767        .005       .004      -1.203
   4      76.700     -17.700        .015       .016      -1.452
   5      66.104      -4.104        .042       .003       -.340
   6      91.006     -15.006        .006       .004      -1.223
   7      84.648       3.352        .006       .000        .272
   8      76.171      24.829        .016       .033       2.048
   9      96.834     -19.834        .012       .015      -1.626
  10      91.006       4.994        .006       .000        .406
  11     103.192      26.808        .024       .060       2.225
  12      85.708      -4.708        .005       .000       -.382
  13      87.827      25.173        .005       .011       2.066
  14      74.051     -12.051        .020       .010       -.988
  15      71.402      -6.402        .026       .004       -.526
```

Figure 9.5
The SCHLRSID file

The first variable, ESTIMATE, is the estimated criterion scores ($\hat{Y}$). The second variable, RESIDUAL, contains the untransformed residual scores ($Y - \hat{Y}$). LEVERAGE and COOK's D values are reported next, followed by studentized deleted residuals (STUDENT). The final three variables, which are not visible in the figure, are the standard error of prediction (SEPRED) and the criterion and predictor values (READ and VIQ).

■ Analyzing Residuals

After you have saved the residuals from a regression analysis, it is often informative to use the plot commands discussed in Chapter 3 to analyze those residuals. Here are some examples you may wish to try using the SCHLRSID file you just produced. (Don't forget to type the USE command to load the SCHLRSID file into MYSTAT.)

- **PLOT RESIDUAL * ESTIMATE** produces a scatterplot of residuals by estimated values.

- **PLOT RESIDUAL * VIQ** produces a scatterplot of residuals by predictor values. This plot is used to assess the nonlinearity of the predictor-criterion relation.

- **TPLOT RESIDUAL** is used to look for a serial pattern in the residuals.

- **STEM RESIDUAL** is useful to look at the shape of the residuals.

- **BOX LEVERAGE** checks for extreme leverage values.

Exercises

1. In Chapter 8, you were asked to plot sextant Dip corrections (DIP) on the *y*-axis against the height of the observer's eye (HEIGHT) variable on the *x*-axis using the file DIPS. You then rank-ordered DIP, saved this file as DIPSRANK, and replotted the variables. Finally, you rank-ordered HEIGHT in the DIPSRANK file and called the new file DIPSRNK2. In this last file, both variables were rank-ordered. You then plotted both ranked variables. In the first plot, you noted that the data formed a smooth curve. In the second plot, the curve was more pronounced. In the final plot you produced a straight line. Calculate the Pearson correlations between the two variables in all three data sets.

    r_{xy} for DIPS _____

    r_{xy} for DIPSRANK _____

    r_{xy} for DIPSRNK2 _____

    Why does the value of the correlation change? If another student says your correlation analysis is inappropriate because these files contain variables with functional relationships, how would you reply?

2. In the file FORECAST are the names of sixteen world cities (CITY$), their recorded high temperatures on July 18, 1991 (RECORDED), and their projected high temperatures for July 19, 1991 (PROJECT). What is the correlation between the temperature measured in a city on one day and the projected temperature for that same city on the following day?

3. Conduct a bivariate regression in which you use VIQ to predict ARITH using the SCHOOL data set. Write a summary statement.

4. Are any outliers identified in the analysis in exercise 3? Write down the case numbers and the MYSTAT indicators if present.

5. Produce a scatterplot with the regression equation for exercise 3.

6. Using the generated regression equation, what would be the expected arithmetic score for a child with a VIQ of 128?

7. In exercise 4 in Chapter 8 you produced a file called DELETE in which you dropped high and low FSIQ scores. Run a regression analysis using this data set in which FSIQ is used to predict READ. Compare these results with those you produced with the full data set earlier. Why are there differences?

8. Use the STATES file to conduct a bivariate regression in which you use winter temperatures to predict population densities. Is there a significant regression equation? Write a summary statement for this experiment.

9. The KARACHI file was introduced in Chapter 7. Calculate the correlation between the variables BACTERIA, LEAD, ARSENIC, CADMIUM, and CYANIDE. Lead is almost universally present in the food. In only twelve samples was no lead found. Eight of those twelve samples were bottled colas. However, arsenic, cadmium, and cyanide are not universally found. When one of these three ingredients is found, how frequently do the other two occur? Some of these correlations are conducted with one or both of the variables measured dichotomously. What might be another name for the correlations?

──────── Notes

[1] You can use MODEL statements to conduct ANOVA calculations. Here are some ANOVA problems and their equivalent MODEL procedures.

Here is a simple one-way ANOVA where GROUP has three levels.

ANOVA Command	MODEL Command
CATEGORY GROUP = 3	CATEGORY GROUP = 3
ANOVA Y	MODEL Y = CONSTANT + GROUP
ESTIMATE	ESTIMATE

Here is a 2 x 3 ANOVA with an A x B interaction term (A * B)

CATEGORY A = 2, B = 3	CATEGORY A = 2, B = 3
ANOVA Y	MODEL Y = CONSTANT + A + B + A * B
ESTIMATE	ESTIMATE

[2] In the top line of output, MYSTAT lists the dependent variable (DEP VAR:) as READ, with 200 cases. Directly to the right, you will see the value .542 labeled "MULTIPLE R." The MODEL and ESTIMATE commands work for multiple regressions as well as bivariate regressions. In multiple regressions the correlation is calculated using several predictors and is called a *multiple R* instead of a simple Pearson correlation (r). However, in this case, because we are only analyzing one predictor, the listing for multiple R is the same as the Pearson r that you found (look at Table 9.1 where VIQ and READ intersect). At the end of this first line, the correlation is squared and labeled SQUARED MULTIPLE R with a value of .294. In some textbooks this is called the *coefficient of determination*. The coefficient of determination tells you the percent of variance in the dependent variable, READ in this case, accounted for by the predictor variable, VIQ in this case. This value is then adjusted and labeled in the second line as the ADJUSTED SQUARED MULTIPLE R, whose value equals .290. This adjusted squared multiple R indicates the percentage of variance that the regression equation would account for if it were used to predict for cases in the population from which this data set was sampled. Next, the STANDARD ERROR OF ESTIMATE is reported. The standard error of estimate is 12.318, which is the standard deviation of the errors in the prediction.

Look back at Figure 9.3. The distance that the regression equation misses the dots by is called the error or the *residual*. The standard error of estimate is the standard deviation of these errors.

The middle section of the output details the regression equation and some associated statistics. Looking under COEFFICIENT, you will note that the CONSTANT is 42.262 and the VIQ is 0.53. You learned that MYSTAT reports the regression equation using the general equation of $y = a + bx$, the constant is the a value and the VIQ coefficient is the b value or slope in the regression equation. These values are also known as *regression coefficients*. If you want to predict another child's READ score and you know his or her VIQ, the following equation is best: READ = 42.262 + 0.530VIQ. If a child has a VIQ of 90, the expected READ score is 89.962.

The symbols used for regression lines change from text to text. In this tutorial a straight line equation will be represented by the general equation $y = a + bx$. In some texts this equation is written $y = bx + c$ or $y = mx + b$. Don't let the changes in abbreviations confuse you. Also, since a regression equation produces expected scores, the general form for the regression equation in this tutorial is $\hat{Y} = aX + b$. The $\hat{Y}$ value indicates a predicted score.

In the next column are the standard errors (predicted STD ERROR) of the coefficients. In the following column are the standard coefficients, labeled "STD COEF." These values are used in a regression equation when all the data values are turned into standard scores by subtracting the mean of each variable from the variable's value and dividing that difference by the standard deviation of the variable. Scores produced by this procedure are called Z scores and typically have means of zero and standard deviations of one. Note that the constant disappears (its value becomes zero), and for bivariate regressions the standardized coefficient for the predictor is the Pearson correlation. In some texts these standardized coefficients are called *beta weights*.

Next to the standardized coefficients are tolerance values. The tolerance value is only useful in multiple regressions. It is a measure of the uniqueness of each predictor in the regression problem. Since there is only a single predictor in a bivariate regression (VIQ), it is unique and always has a tolerance value of 1.0 (this value is written in scientific notation in Table 9.4). The t values in the last column are inferential tests (*t*-tests) that determine if the coefficients for the constant and predictors are different from zero. Rarely will the constant ever be tested, so that t value is infrequently used. In a bivariate regression there is only one predictor and the t value reported is simply the square root of the F value reported in the ANALYSIS OF VARIANCE table at the bottom. In bivariate regression, both the t and the F value give the same information.

10

An Introduction to Analysis of Covariance

In Chapter 7, we discussed analysis of variance (ANOVA) techniques. ANOVA is used to compare the means of more than two groups, and the effect of the categorical variable on the values of the dependent variable is evaluated. In Chapter 9, we discussed correlation and regression techniques that evaluate the effect of one or more quantitative independent variables on the values of a single continuous dependent variable. In this chapter, we will merge the two techniques. We will analyze data that contain both categorical and quantitative independent variables, and we will separate the effects of these different varieties of independent variables. This analysis technique is called *analysis of covariance,* or *ANCOVA.* While analysis of covariance (ANCOVA) is sporadically covered in beginning courses, understanding the general idea behind ANCOVA and its ability to increase the power of your statistical decisions is important.

Objectives

At the end of this tutorial you should be able to

- Decide when to use ANCOVA
- Know the two major uses of ANCOVA
- Acknowledge which of those uses is more often justified
- Direct MYSTAT to conduct ANCOVA

■ Determining Whether ANCOVA Is Appropriate

With ANCOVA techniques, unlike ANOVA or regression techniques, you have two different types of independent variables. At least one categorical and one quantitative independent variable are present. Figure 10.1 illustrates the decision model for determining when to use ANCOVA.

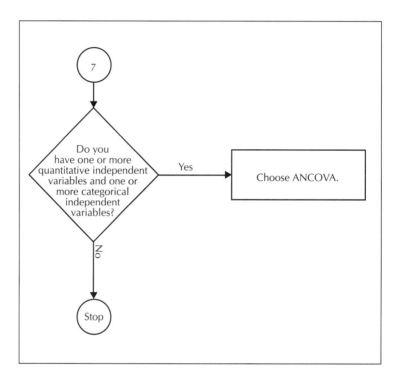

Figure 10.1
Decision model for ANCOVA

■ What Is ANCOVA?

Analysis of covariance, or ANCOVA, is a combination of regression and ANOVA techniques. In ANCOVA, the quantitative independent variable is called the *covariate* and the categorical independent variable is called the *factor*. Usually the covariate is used in a regression analysis to correct or adjust the dependent scores and then a regular analysis of variance is conducted using these corrected scores. Under certain circumstances researchers gain a great deal of power using these analyses. (Remember, the power of a statistical test is its ability to detect differences in the dependent variable.)

ANCOVA (indeed, even ANOVA) is also taught as an elaboration of regression analysis in which both quantitative and categorical predictors are used. Using this approach, it is logical for a researcher to be interested in the effects of the quantitative variable while controlling for the effects of the categorical variable. This ANCOVA approach is often used in research investigating bias (for example, to tell whether a test is biased by race or sex, whether an employer discriminates by age, and so on).

In ANCOVA, the covariate may be included in the design for two purposes. The first purpose is to statistically equate groups that are not similar. There are real problems with this practice, and almost all textbooks caution against using ANCOVA for group adjustment. When reading papers in which ANCOVA is used for this purpose, you should interpret the results with caution. The reasons for these interpretation difficulties are beyond the scope of this tutorial, but they are discussed in detail in advanced statistics texts.

The second purpose for using ANCOVA is to increase the power of the statistical analysis. If there is a covariate (quantitative variable) unrelated to the grouping condition (categorical variable) but related to the dependent variable, including the covariate in the analysis may dramatically increase the power of the statistical test. This is easily accomplished if the investigator can randomly assign subjects to different treatment groups.

For example, if you are a teacher interested in how a new reading curriculum affects reading scores, you might give the children a reading pretest (the covariate), randomly assign the children to different treatment groups, and after using the new curriculum give a reading posttest. The reading pretest and posttest scores will be related. You expect children with high scores on the reading pretest to have high scores on the reading posttest and children with low scores on the reading pretest to have low scores on the reading posttest. There is a correlation between the level at which children read today and the level at which they read in the future. In Chapter 9, you learned that the square of the correlation coefficient (the coefficient of determination) indicates how much variance in the dependent variable is accounted for by the predictor variable. By using the pretest results to predict the posttest results, some of the variance in the dependent variable can be accounted for. Because the covariate (pretest) and the grouping variable (factor) have no relationship to each other, the variance accounted for in the dependent variable by the covariate will have no relationship to the categorical independent variable. Remember from Chapter 7 that variance not related to the treatment is error variance. Therefore, if the variance in the dependent variable accounted for by the covariate is statistically removed, you decrease the error variance. Also, since the error variance is in the denominator of the F statistic, if you decrease the denominator of a fraction, the value of the fraction increases. Therefore by including a covariate, you will increase the value of the fraction (the F value). The only problem is that you lose one degree of freedom for the covariate. So the amount of error variance removed by the covariate must compensate for the degree of freedom lost. Following is an example that illustrates the increase in power one may get using ANCOVA instead of ANOVA.

The Problem

You are a research coordinator for the American Nursing Association. You are interested in a new approach to teaching surgical assistance in a cardiac valve replacement. You give the nurses a pretest on their knowledge of this material (PRETEST). They are randomly assigned to two different curricular approaches

(CURRICUL) and to two different instructors (TEACHER). After instruction, you give a posttest (POSTTEST). You want to find out whether the curriculum, the teacher, or a combination (this is called an *interaction*) makes a difference in the posttest scores. These data are stored in the CARDIAC file on your data disk.

The ANOVA Solution

The ANOVA results for this problem are presented first. These results will then be compared to the ANCOVA results. To analyze the CARDIAC data using ANOVA, you want to detect whether there are differences in the dependent variable (POSTTEST) using the categorical variables CURRICUL and TEACHER. Each independent variable has two different levels. You will be conducting a 2 x 2 ANOVA.

Use the same six-step solution that you used to solve the problem using the SCHIZOPH file on pages 88 through 90 in Chapter 7, but to calculate the statistic follow these directions:

1. Type **USE B:CARDIAC** and press **[Enter]** to load the data.

   Note that there is a variable for the pretest (PRETEST), which is the covariate or quantitative independent variable. The dependent variable is POSTTEST, and the two variables for the curriculum type (CURRICUL) and the teacher (TEACHER) are the categorical independent variables.

2. Press **[Enter]** to return to the main command screen.

3. Type **CATEGORY CURRICUL = 2, TEACHER = 2** and press **[Enter]** to choose both CURRICUL and TEACHER as the independent variables with two levels.

4. Type **ANOVA POSTTEST** and press **[Enter]** to choose POSTTEST as the dependent variable.

5. Type **ESTIMATE** and press **[Enter]** to calculate the statistic.

   Your results should look like those presented in Table 10.1.

DEP VAR: POSTTEST   N: 40   MULTIPLE R: .646   SQUARED MULTIPLE R: .417

ANALYSIS OF VARIANCE

SOURCE	SUM-OF-SQUARES	DF	MEAN-SQUARE	F-RATIO	P
CURRICUL	422.500	1	422.500	21.423	0.000
TEACHER	22.500	1	22.500	1.141	0.293
CURRICUL*TEACHER	62.500	1	62.500	3.169	0.083
ERROR	710.000	36	19.722		

Table 10.1
ANOVA results for the Cardiac data

If the alpha level is set at .05, there is one significant finding using ANOVA. The curriculum does make a difference in the nurses' scores. Note that the error sum of squares is 710 with 36 degrees of freedom, providing a mean-square error of 19.722.

The ANCOVA Solution

With ANCOVA, you use the covariate (PRETEST), which is unrelated to the categorical independent variables (CURRICUL and TEACHER). The covariate is used to reduce the sum-of-squares error term. As mentioned earlier, one degree of freedom will be used by the covariate, so the sum-of-squares accounted for by the covariate must be worth the degree of freedom lost to the error term. Remember from Chapter 7 that if you divide the sum-of-squares by the degrees of freedom (DF) you will obtain the mean-square value. If the mean-square error is made smaller, the F statistic will be larger, and the power of the test will be increased. You will use the six-step solution.

1. State the null and alternative hypotheses.

In ANCOVA there may be several sets of null and alternative hypotheses. In this problem, you will have a null and alternative hypothesis for each of the independent variables (both quantitative and categorical) and for the interaction between the categorical predictors. ANCOVA hypotheses are stated without symbols.

H_{O_1}: There is no curricular effect on posttest scores.

H_{1_1}: H_{O_1} is false.

H_{O_2}: There is no teacher effect on posttest scores.

H_{1_2}: H_{O_2} is false.

H_{O_3}: There is no interaction effect between curriculum and teacher on posttest scores.

H_{1_3}: H_{O_3} is false.

Notice that these are the same hypotheses as in the ANOVA. We are using ANCOVA to increase the power of each of these tests.

2. Set the alpha level.

We'll use the same alpha level as we did for the ANOVA solution, .05.

3. Collect the data.

The data for this problem are found in the file CARDIAC. To list the variable names of the file currently loaded, use the **NAMES** command.

1. Type **NAMES** and press [**Enter**].
2. Press [**Enter**] to return to the main command screen.

4. Calculate the statistic.

To change from an ANOVA to an ANCOVA, you need to indicate which variables are to be treated as covariates with the COVARIATE command. The general form of the COVARIATE command is

COVARIATE = <VAR1>

Do the following:

1. Type **CATEGORY CURRICUL = 2, TEACHER = 2** and press **[Enter]**.
2. Type **COVARIATE = PRETEST** and press **[Enter]** to tell MYSTAT that the pretest values are to be used as the quantitative independent variable, or covariate.
3. Type **ANOVA POSTTEST** and press **[Enter]**.
4. Type **ESTIMATE** and press **[Enter]**.[1]

   The ANCOVA output is shown in Table 10.2.

DEP VAR: POSTTEST   N: 40   MULTIPLE R: .845   SQUARED MULTIPLE R: .714

ANALYSIS OF VARIANCE

SOURCE	SUM-OF-SQUARES	DF	MEAN-SQUARE	F-RATIO	P
CURRICUL	422.500	1	422.500	42.484	0.000
TEACHER	22.500	1	22.500	2.262	0.142
CURRICUL*TEACHER	62.500	1	62.500	6.285	0.017
PRETEST	361.929	1	361.929	36.394	0.000
ERROR	348.071	35	9.945		

Table 10.2
ANCOVA results for the Cardiac data

In the ANCOVA results, the interaction (CURRICULUM * TEACHER) is significant: $p = 0.017$. This indicates that the curriculum and the teacher together affect the dependent variable when the pretest is controlled. When a single categorical independent variable is referred to in either ANOVA or ANCOVA, it is sometimes referred to as a *main effect* to distinguish it from interactions. The first main effect, CURRICUL, is significant ($p < 0.0005$). The second main effect, TEACHER, is nonsignificant ($p = 0.142$). Finally, the pretest is a significant predictor (or covariate) of the scores ($p < 0.0005$), although this statistical significance is not necessary for the ANCOVA to be valid, and it is not one of our hypotheses.

WARNING: *With a significant interaction you should be careful about interpreting significant main effects.*

Let's compare the two summary tables starting with the ERROR line. In the ANOVA table, the error sum-of-squares is 710, while in the ANCOVA table, the error sum-of-squares is 348.071. The ANCOVA error sum-of-squares is smaller. Why? In the ANCOVA table there is an extra line for the effect of the covariate (PRETEST). The covariate accounts for 361.929 sum-of-squares. If you add 348.071 and 361.929 together, the result is 710. The sum-of-squares due to the covariate were removed from the ANOVA error term. The degrees of freedom for the error term in ANOVA are 36, giving a mean-square of 19.722. In ANCOVA the degrees of freedom are 35 (you lost one for the covariate), but the mean-square is 9.945, which is considerably smaller than in the ANOVA results. Now look at the curriculum by teacher interaction (CURRICUL * TEACHER). In both the ANOVA and ANCOVA tables, the sum-of-squares accounted for by the interaction is 62.5. In both Table 10.1 and Table 10.2, the degrees of freedom for the interaction are 1 and the mean-square is 62.5. In the ANOVA table, the *F* value for the interaction is 3.169, which is nonsignificant ($p = 0.083$). In the ANCOVA table, the *F* value is 6.285. This value is considerably larger because the *F* value is found by dividing the mean-square of the source by the mean-square error. Since the mean-square error is reduced in ANCOVA, the *F* value is larger and the test is more powerful. In ANCOVA, the interaction is significant. Because of its lack of power, the ANOVA procedure was unable to detect this interaction.

This example clearly demonstrates the increase in statistical power available using ANCOVA procedures. Notice how all the other main effect *F* values have increased.

5. Decide whether or not to reject the null hypothesis.

In this case you would reject the first and third null hypotheses and fail to reject the second. If you had generated a fourth hypothesis concerning the covariate (e.g. H_{O4}: There is no relationship between PRETEST and POSTTEST), this would be rejected also.

6. Write a summary statement.

For this problem, the researcher might report: "The curriculums made a difference in the nurses' scores on the cardiac valve replacement exam (F = 42.484, df = 1, 35, p < 0.0005). There was no significant difference in the test scores, which were dependent upon which teacher the students had (F = 2.262, df = 1, 35, p = 0.142). However, there was a significant curriculum-by-teacher interaction (F = 6.285, df = 1, 35 p = 0.017). Finally, the relationship between the pretest and posttest scores was significant, controlling for both curriculum and teacher effects (F = 36.394, df = 1, 35, p < 0.0005)."

■ Assumptions of ANCOVA

ANCOVA has assumptions that are similar to those for ANOVA. The residuals are supposed to be random samples from a population of errors that are normally distributed with the same variance in every group and a mean of zero in every group. You can save and examine the residuals to see how plausible these assumptions are for your data.

There is an additional set of assumptions that is especially important for ANCOVA, however. Since we are making a common adjustment for all groups, the covariate should be related linearly to the dependent measure in the same way for every group. Thus, for example, if we plotted a scatterplot between POSTTEST and PRETEST for every group, all four scatterplots should look similar and the regression lines predicting POSTTEST from PRETEST should have similar slopes. This is often called the "homogeneity of slopes" assumption.

Exercises

1. Conduct an ANCOVA using the SCHOOL data. Use MDT as the independent variable (factor), READ as the dependent variable, and VIQ as the covariate. If a group is reading disabled, it should have significantly different scores on the reading test than other groups. Reading disabled children are supposed to be underachieving in reading. This difference should show when verbal intelligence is used as a covariate. Write a summary statement. You will need to delete MDT group 7, as there is only a single subject in this group. (A group can't be formed using a single case because the variance in the group is zero.) Without group 7 there are six remaining groups.

2. Use the STATES data. In Chapter 6, exercise 3, you created a variable that separated the eastern states from the western states. Determine whether there are significant differences between the population densities of these two groups using winter temperature as a covariate. Write a summary statement for your results.

3. Use data listed in your text or assigned by your instructor. Conduct the ANCOVA and write summary statements.

Notes

[1] As the note at the end of chapter 9 indicates, MODEL and ESTIMATE commands can be used to conduct ANOVA commands. They can also be used to conduct ANCOVA analyses.

Here are equivalent commands for a 2 x 3 ANCOVA.

ANOVA Command	**MODEL Command**
CATEGORY A = 2, B = 3	CATEGORY A = 2, B = 3
COVARIATE = X	MODEL Y = CONSTANT + X + A + B + A * B
ANOVA Y	ESTIMATE
ESTIMATE	

Here is an ANCOVA analysis that includes factor (A, B) by covariate (X) interactions. This analysis might be used to test the homogeneity of regression slopes assumption.

CATEGORY A = 2, B = 3
MODEL Y = CONSTANT + X + A + B + A * B + A * X + B * X + A * B * X
ESTIMATE

11

Nonparametric Statistical Tests

Until now, we have discussed and learned how to calculate *parametric statistics*. In parametric statistics, the symbols used when writing the null and alternative hypotheses are population parameters. These parameters completely specify the location and shape of a normal distribution.

Nonparametric statistics are often termed distribution-free tests. They do not assume that a population distribution must be specified by parameters. Consequently, these procedures do not use population parameters in their null or alternative hypotheses.

A complete discussion of nonparametric statistics is beyond the scope of this book. Additional information on nonparametric statistics can be found in S. Siegel (1956) and L. A. Marascuilo and M. McSweeney (1977).

Objectives

After completing this tutorial you should be able to

■ Determine which nonparametric procedure is appropriate

■ Calculate a one-way chi-square

■ Calculate a two-way chi-square

■ Calculate a Sign test

■ Calculate a Wilcoxon signed-rank test

■ Calculate a Friedman nonparametric analysis of variance

■ Determining Which Nonparametric Procedure to Use

You use the *one-way chi-square* test when both the independent and dependent variables are categorical. Use this statistic when you have a single independent variable and you want to compare the number of times (frequency) a category or group defined by the independent variable occurs with the number of times you expect it to occur. Usually, the expected frequencies are equal for each group, so this statistic evaluates whether the frequencies for the groups are different.

You use *two-way chi-square* tests when there are two independent categorical variables and a single categorical dependent variable. Like the one-way chi-square, the two-way chi-square determines whether the number of times a category occurs is different from what is expected. When the expected frequencies in a two-way chi-square are determined, it is assumed that the two independent variables are unrelated. Therefore, if the observed frequencies for the groups are different from the expected frequencies, there is a relationship between the independent variables.

The *Sign* test is the nonparametric equivalent of the dependent *t*-test (see Chapter 6). The independent variable is categorical and consists of two levels like the dependent *t*-test. The dependent variable is assumed to be quantitative, but it can't be measured using a quantitative scale, so a categorical scale is substituted. This often happens when an appropriate test is unavailable, but an observer can count when the dependent variable takes one of two values (usually present or absent). The Sign test is inappropriate if the dependent variable has more than two values. Variables that take only two different values are called *binary* variables. Using the Sign test you want to know if there are differences between the two groups on the binary variable.

The *Wilcoxon* test is also a nonparametric alternative to the dependent *t*-test. The independent variable is categorical and has two levels. The dependent variable is categorical, but the dependent variable values have an order. The numbers indicate a quantity, but you only know if a case has more or less of the variable, not how much more or less. Positions in a race are good examples of this type of variable. If you know that Marsha took first place (1), Mary took second place (2), and Maria took third (3), you know that Marsha did better than Mary, but you don't know how much better. The race may have required a photo finish or Marsha may have lapped Mary. You also know that Mary did better than Maria. This test is sometimes called the *Wilcoxon signed ranks test*—the dependent variable is a rank. You want to know if there is a difference in the ranks on the dependent variable between the two related groups.

If you measure a subject two or more times, you are doing a *repeated measures analysis*. You use the *Friedman analysis of variance* test if you have a categorical independent variable (there may be more than one independent variable and each independent variable may have more than two levels) and a rank-order dependent variable that is measured more than twice. Use this test to determine whether the dependent variable differs across the groups formed by the independent variable(s).

Figure 11.1 illustrates the decision model for choosing among these nonparametric tests.

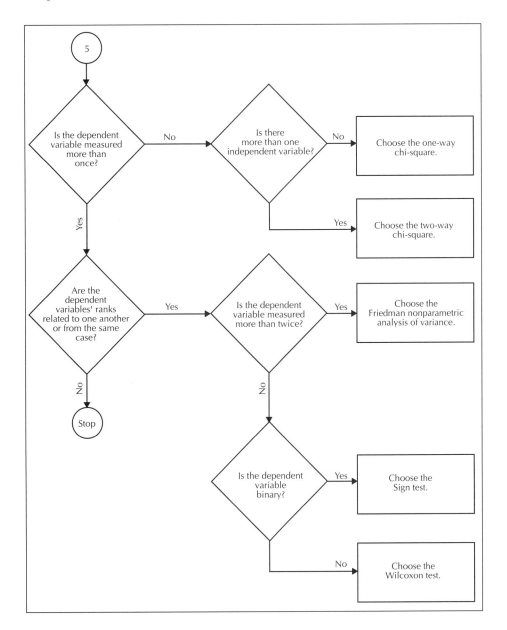

Figure 11.1
Decision model for
nonparametric tests

■ The TABULATE command

The TABULATE command produces frequency and *n*-way crosstabulation tables. For two-way tables TABULATE provides chi-square test statistics, association coefficients, and other measures of association along with their asymptotic standard errors. The general form of the TABULATE command is:

TABULATE <VAR1> [* <VAR2> * <...>][/LIST, FREQUENCY, PERCENT, ROWPCT, COLPCT, MISS]

You can display frequencies, row percents, column percents, or cell percents in the table. Frequencies can be in either LIST or default formats. If the LIST option is typed, neither the chi-square statistics nor the measures of association are calculated. The MISS option allows you to ignore missing data. Following are some valid TABULATE commands:

■ **TABULATE AGE**   produces a one-way table.

■ **TABULATE AGE/LIST**   produces a one-way table in list format.

■ **TABULATE AGE * SEX**   produces a two-way table with the associated statistics.

■ **TABULATE AGE * SEX * STATE$**   produces a three-way table.

■ **TABULATE AGE, SEX * STATE$**   produces two two-way tables. Note that a comma separates the first two variables instead of an asterisk.

■ **TABULATE AGE * SEX/FREQ, PERCENT**   produces two two-way tables. The first table contains cell frequencies, the second table's cells contain percents. These two tables are then followed by the associated statistics.

■ **TABULATE AGE/MISS**   produces a one-way table and ignores missing data.

■ One-way Chi-Square

As noted, a one-way or one-variable chi-square (χ^2) test (also called the *goodness-of-fit test*) compares a set of observed frequencies (O) with an expected set of frequencies (E). The observed frequencies for each group are the number of sample cases in each group. The expected frequencies are the number of cases you expect each group to have given the sample size. Expected frequencies are determined in two ways. First, you might have a theoretical reason to expect a certain percentage of the frequencies to be in a particular category. For example, genetic theories for dominant and recessive traits lead researchers to conclude that if the theory is correct, a certain percentage of offspring should have the dominant trait while the remainder would possess the recessive trait. Second, in the absence of a theoretical expectation, you expect random assignment to the categories. This is the procedure MYSTAT uses to assign expected frequencies. Using random assignment, each category is expected to have the same number of frequencies. If the observed frequencies are distributed differently from the expected frequencies, the population distributions for the groups are presumed to be different.

The Problem   You are a market researcher trying to decide which of four packages created for your client's product is best. You present the four different packages to 100 consumers and ask them to make a choice concerning which packaging they prefer. You code subjects with a 1 if they choose package #1, a 2 if they choose package #2, and so on. If there are no differences in preference for the four packages, you assume that each package receives an equal number of choices. Twenty-five people would choose package #1 as best, twenty-five people would choose package #2 as best, twenty-five people would choose package #3 as best, and twenty-five people would choose package #4 as best. Twenty-five is the expected frequency for each group (E). To compare the observed frequencies (O) with the expected frequencies (E), you calculate a chi-square statistic.

The Solution

As before, use six steps.

1. Write the null and alternative hypotheses.

The null hypothesis in chi-square states that the distribution of observed frequencies is identical to the distribution of expected frequencies.

H_0: O = E

H_1: O $\neq$ E

2. Set the alpha level.

We'll use .05 again.

3. Collect the data.

The data for this problem are found in the file PACKAGE.

4. Calculate the statistic.

To calculate the chi-square, do the following:

1. Type **USE B:PACKAGE** and press [**Enter**].

   Note that there is a single variable (CHOICE) that indicates which of the four packages was preferred.

2. Press [**Enter**] to return to the main command screen.

3. Type **TABULATE** and press [**Enter**] to produce a table of observed frequencies for each numeric variable in the data.

   Table 11.1 indicates that there were 9 people who preferred package #1, 58 who preferred package #2, 19 who preferred the package #3, and 14 who preferred package #4.

TABLE OF VALUES FOR CHOICE
FREQUENCIES

	1.000	2.000	3.000	4.000	TOTAL
	9	58	19	14	100

TEST STATISTIC	VALUE	DF	PROB
PEARSON CHI-SQUARE	60.080	3	.000
LIKELIHOOD RATIO CHI-SQUARE	52.569	3	.000

Table 11.1
Tabular results for the
CHOICE variable

MYSTAT prints the Pearson chi-square below the table, assuming equal expected frequencies for each package. The formula for a chi-square is:

$$\chi^2 = \sum_{cells} \frac{(O - E)^2}{E}$$

The Σ stands for "add what follows." Cells is used as a subscript to remind you to add what follows for all the cells or groups defined by the independent variable. In many textbooks you will find subscripts and superscripts around the summation sign, which indicate that you add together each of the values found for each group. In this case, there are four groups or cells, one for each preferred package. Remembering that for each of the cells the expected frequency (E) is 100/4 = 25, you will take each observed frequency (found in Table 11.2), subtract 25 from that value, square the result, and divide by 25, and add the values generated for each cell together to calculate the chi-square.

Your calculated χ^2 value equals 60.08, which is what MYSTAT prints.

5. Decide whether or not to reject the null hypothesis.

The critical value for chi-square marks the boundaries for rejecting the null hypothesis similar to the critical z value discussed in Chapter 5. Unlike the z critical values, χ^2 critical values change for different degrees of freedom. If your obtained χ^2 is larger than the critical χ^2, then you reject the null hypothesis. Also, look at the probability. If it is less than the alpha level, you can reject the null hypothesis. Figure 11.2 illustrates this decision.

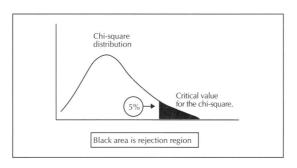

Figure 11.2
Critical χ^2 values

In this case you reject the null hypothesis. Your calculated χ^2 value of 60.08 has a $p < .0005$ as shown in Table 11.2.

6. Write a summary statement.

For this problem, the researcher might report: "The different packages were not equally preferred. Subjects showed a clear preference for package #2 ($\chi^2 = 60.08$, df = 3, p < .0005)."

■ Two-way Chi-Square

Another name for the two-way chi-square is the *chi-square test of independence*. When the dependent variable is measured at the categorical level and the researcher has an experimental situation with two independent variables, each with at least two categories, the two-way chi-square test is appropriate.

The Problem

You are a researcher for the Department of Defense interested in determining whether there is a relationship between the social class of military volunteers and the branch of service for which they volunteer. You collect data for a random sample of 100 recruits. The two variables are CHOICE (Army = 1, Navy = 2, Air Force = 3, and Marines = 4), and STATUS (upper class = 1, middle class = 2, lower class = 3).

The Solution

1. State the null and alternative hypotheses.

More often than not you will see this test's null and alternative hypotheses stated without symbols.

H_0: There is no relationship between the branch of service and the social class of military recruits.

H_1: There is a relationship between the branch of service and the social class of military recruits.

Obviously, the null and alternative hypothesis could be stated symbolically as was done for the goodness-of-fit test.

H_0: O = E

H_1: O $\neq$ E

2. Set the alpha level.

Set the alpha level to .05 again. MYSTAT will calculate the chi-square statistic, its probability, and other associated values.

3. Collect the data.

The data for this problem are found in the file MILITARY.

4. Calculate the statistic.

From the main command screen,

>   1. Type **USE B:MILITARY** and press [**Enter**] twice to load the data.
>   2. Type **TABULATE CHOICE * STATUS**.
>
>   Note the asterisk in the TABULATE command, which indicates that the groups are formed by crossing the two variables. When the two variables are crossed, the groups formed are sometimes referred to as cells. For example, if the first variable has four different values (levels) and the second variable has three different values, twelve groups or cells will be formed. The table size has a size of 4 x 3. This is illustrated in Figure 11.3.

		Three levels of the second variable		
		1.000	2.000	3.000
	1.000	cell 1	cell 2	cell 3
Four levels of the first variable	2.000	cell 4	cell 5	cell 6
	3.000	cell 7	cell 8	cell 9
	4.000	cell 10	cell 11	cell 12

Figure 11.3
A table of size 4 x 3

>   3. Press [**Enter**] to calculate the chi-square.
>
>   You should see the output shown in Table 11.2.

5. Decide whether or not to reject the null hypothesis.

MYSTAT automatically provides the table and calculates two different chi square statistics. The likelihood ratio chi-square is an alternative to the more familiar Pearson chi-square and is used in log-linear analyses, which are beyond the scope of this book. The information you need to reject the null hypothesis is contained in the Pearson chi-square row. In this case you reject the null hypothesis. Your calculated χ^2 of 22.48 has a probability of .001, which is lower than the .05 alpha level.

6. Write a summary statement.

For this problem, the researcher might report: "There is a significant relationship between the type of military service for which recruits volunteer and the recruits' social status (χ^2 = 22.48, df = 6, p = .001)."

TABLE OF CHOICE (ROWS) BY STATUS (COLUMNS)
FREQUENCIES

	1.000	2.000	3.000	TOTAL
1.000	20	19	3	42
2.000	6	14	5	25
3.000	2	5	10	17
4.000	6	5	5	16
Total	34	43	23	100

TEST STATISTIC	VALUE	DF	PROB
PEARSON CHI-SQUARE	22.480	6	.001
LIKELIHOOD RATIO CHI-SQUARE	21.993	6	.001

COEFFICIENT	VALUE	ASYMPTOTIC STD ERROR
PHI	.4741	
CRAMER V	.3353	
CONTINGENCY	.4284	
GOODMAN-KRUSKAL GAMMA	.3951	.11841
KENDALL TAU-B	.2790	.08642
STUART TAU-C	.2829	.08797
SPEARMAN RHO	.3097	.09626
SOMERS D (COLUMN DEPENDENT)	.2669	.08404
LAMBDA (COLUMN DEPENDENT)	.1228	.13247
UNCERTAINTY (COLUMN DEPENDENT)	.1030	.04125

Table 11.2
TABULATE output
for a two-way
chi-square

Other Measures of Association

In Table 11.2, other measures of association are reported in addition to the Pearson and likelihood chi-squares. The Pearson chi-square indicates whether the variables are associated; these other measures of association indicate the strength of the association. While a complete description of each measure of association is beyond the scope of this book, a brief summary follows.

Phi, Cramer V, and the contingency coefficient all quantify the degree of association tested by the Pearson chi-square. Indeed, their formulas are all based directly on the chi-square formula. For 2 x 2 tables (tables with two groups on both variables), *phi* is identical to a Pearson correlation calculated on binary variables and has values between −1.0 and +1.0 (see Chapter 9). When *phi* is calculated for tables larger than 2 x 2, it does not have an upper limit and is quite difficult to interpret.

Cramer V is a slight modification of *phi.* Cramer V can take values from 0 to +1. A large value for Cramer V signifies that a high association exists between the variables.

The contingency coefficient is also derived from the chi-square value. Its values range from zero to some upper limit. The upper limit of the contingency coefficient depends upon the table size, so contingency coefficients should only be compared between tables of the same size. The larger the contingency coefficient, the stronger the relationship between the variables.

Goodman-Kruskal *gamma*, Kendall *tau*-B, Stuart *tau*-C, Somers D, and Spearman *rho* are appropriate when both categorical variables are ranks. Spearman *rho* (or rank-order correlation) was discussed in Chapter 9. *Tau*-B, *tau*-C, *gamma*, and Somers D all use information about the ordering of the variables by considering every pair of values in the data set. The first four measures primarily differ from one another in how ties in the rankings are treated. Because Spearman's *rho* is so well known, it is the usual measure of association reported when the variables are ranks.

Lambda and the uncertainty coefficient are measures of association when the variables are measured at the categorical level. *Lambda* measures the percent of improvement in your ability to predict the value of the dependent variable once you know the value of the independent variable. Obviously, there could be two *lambdas* calculated depending on which of the two variables is considered the dependent variable. MYSTAT always calculates *lambda* using the column variable as the dependent variable.

The uncertainty coefficient is similar to the *lambda* coefficient. It measures the proportion by which uncertainty is reduced in the dependent variable once you know the value of the independent variable. Both *lambdas* and the uncertainty coefficients have values that range from 0 (no improvement) to 1 when perfect predictions are possible.

■ Other Nonparametrics

Other nonparametric statistics available through MYSTAT are the Sign test, the Wilcoxon signed-rank test, and the Friedman nonparametric analysis of variance. All three tests are used with dependent (correlated) samples. The Sign test and the Wilcoxon signed-rank test are both nonparametric equivalents of the dependent *t*-test (see Chapter 6). If there are more than two dependent samples or you wish to test the same subject under more than two conditions, use the Friedman analysis of variance test (see Figure 11.1).

Sign Test

The Sign test is performed on a set of difference scores when only the sign of the differences is retained to compute the statistic. You obtain either positive or negative differences, then the number of positive differences and the number of negative differences are summed. The significance of the difference between these two numbers is tested by the Sign test.

The SIGN command computes a Sign test on all pairs of specified variables. The Sign test in MYSTAT omits differences of zero. The general form of the SIGN command is

SIGN [<VAR1>, <VAR2>, <...>]

If you don't type any variable names after the SIGN command, a Sign test is conducted on all quantitative variables.

The Problem

The Sunfish sailing class association conducts one-design races across the world. In one-design races a person should be able to take a boat directly out of the box, rig it, and win with it. Over the years, competitors have made many modifications to their Sunfishes. One that appears to have made a distinct difference is a modification to the wooden daggerboard. Racers have strengthened the daggerboard by adding fiberglass to the outside. This increased area of the daggerboard prevents the boat from slipping sideways through the water and allows it to sail better upwind. No sailor can win a competition with a boat directly from the manufacturer. The winners are often not the best sailors but the best at laying up fiberglass.

The class association wants to market a new fiberglass board and make it the only daggerboard (no modifications) acceptable for racing. They know that the class will not accept the new board unless it is better than the old board. Eighty sailors who attend the North American Championship regatta are paired using their overall positions during the five-day event to produce 40 matched pairs. One person in each pair is randomly provided the new fiberglass board, while the partner uses the wooden board. If times were kept in the races (a quantitative variable), they would not be comparable because of different wind conditions in the races. However, it is easy to determine which member of a pair wins and which loses (a binary variable). If significantly more people win with the new board, the board will have increased their speed. Since the pairs are matched on sailing performance and the boards are randomly assigned, you would expect that an equal number of winners would have wooden and fiberglass daggerboards if the board did not affect boat speed.

The Solution

1. State the null and alternative hypotheses.

Often you see the Sign test's null and alternative hypotheses stated without symbols. You might see the null and alternative hypotheses written as follows:

H_0: There is no difference in the daggerboards.

H_1: The fiberglass daggerboard is better than the wooden daggerboard.

Because the Sign test deals only with the signs of the differences (either positive or negative), another way of expressing the null hypothesis is to say that the Sign of any difference is just as likely to be positive as it is to be negative.

2. Set the alpha level.

We'll set the alpha level to .05 and MYSTAT will calculate the Sign test's value and its probability.

3. Collect the data.

The data for this problem are found in the file SUNFISH. When you open the data set, note that there are two variables. The first variable is WOOD: if the wooden daggerboard won, a 1 is recorded; if the wooden daggerboard came in second, a 2 is recorded. The second variable, FIBER, measures the position of the boat with the fiberglass board.

4. Calculate the statistic.

1. Type **USE B:SUNFISH** and press [**Enter**] twice.

2. Type **SIGN** and press [**Enter**] to conduct a Sign test on all quantitative variables.

You will receive the output shown in Table 11.3.

SIGN TEST RESULTS
COUNTS OF DIFFERENCES (ROW VARIABLE GREATER THAN COLUMN)

	WOOD	**FIBER**
WOOD	0	30
FIBER	10	0

TWO-SIDED PROBABILITIES FOR EACH PAIR OF VARIABLES

	WOOD	**FIBER**
WOOD	1.000	
FIBER	.003	1.000

Table 11.3
Sign test results

The first table indicates how many times the row variable had a greater value than the column variable. Note that this table is titled "Counts of differences." In the first row you see that the wooden daggerboard (WOOD) had a greater value than the fiberglass board thirty times. This means that the boat with the fiberglass daggerboard beat the boat with the wooden daggerboard thirty times. The second row indicates that the boat with the wooden daggerboard beat the other boat ten times. In the second table, the two-tailed probability of this event is calculated. The *p* value is .003.

5. Decide whether or not to reject the null hypothesis.

Since this was a one-tailed test (look at H_1), you can divide the *p* value in half. Because .0015 is less than the alpha level, reject the null hypothesis.

6. Write a summary statement.

For this problem, the researcher might report: "There was an improvement in finishing positions when using the fiberglass board (Sign test = 30, p < .05)."

When the number of pairs is greater than twenty-five, the null hypothesis may be tested with a *z*-statistic. The formula for the *z*-statistic in this case is

$$z = \frac{(x \pm .5) - \frac{1}{2}n}{\frac{1}{2}\sqrt{n}}$$

where x is the smaller of the two counts of differences and n is the number of untied pairs. In this problem the number of fewer signs is 10, and n is 40. The rule for choosing either $x + .5$ or $x - .5$ is: Use $x + .5$ when $x < .5n$; use $x - .5$ when $x > .5n$. In this case $.5n = 20$, so you will use $x + .5$. Filling in the equation gives

$$z = \frac{10.5 - 20}{3.16227766} = -3.00.$$

You can then consult the normal distribution table in your text or use the ZCF function in MYSTAT (see Chapter 5) to determine if the results are significant. If you use the ZCF function you will note that a z value of -3.00 has a probability of .001.

The Wilcoxon Matched-Pairs Signed Ranks Test

The Wilcoxon matched-pairs signed ranks test, or Wilcoxon for short, is similar to the Sign test, but it is calculated on ranked variables. Because more information is contained in ranked dependent variables than the binary dependent variable of the Sign test, the Wilcoxon is a more powerful test than the Sign test.

The WILCOXON command test in MYSTAT omits differences of zero, averages tied ranks, and adjusts for ties. The general form of the command is:

WILCOXON [<VAR1>, <VAR2>, <...>]

If no variable names are typed after the WILCOXON command, a Wilcoxon test is conducted on all quantitative variables.

The Problem

The Sunfish data used for the Sign test would take a long time to produce. Each matched pair of sailors races and the winner and loser must be recorded. In effect, you would have to run forty matched races. But if all the boats could be started at one time and their finishing positions noted, then the data is a natural for the Wilcoxon procedure because the dependent variable is of rank order. Eighty boats on a starting line would be too many, however, so only the first 20 matched pairs (40 boats) are selected to participate in this experiment.

The Solution

Steps 1-2 are identical to those in the Sign test and are not repeated here.

3. Collect the data.

The 20 matched pairs start the race simultaneously, and the race committee notes the position of each boat as it crosses the finish line. The data are contained in the SUNFISH2 file.

4. Calculate the statistic.

1. Type **USE B:SUNFISH2** and press [Enter] twice.
2. Type **WILCOXON** and press [Enter].

The results are presented in Table 11.4.

WILCOXON SIGNED RANKS TEST RESULTS

COUNTS OF DIFFERENCES (ROW VARIABLE GREATER THAN COLUMN)

	WOOD	**FIBER**
WOOD	0	12
FIBER	8	0

Z = (SUM OF SIGNED RANKS)/SQUARE ROOT (SUM OF SQUARED RANKS)

	WOOD	**FIBER**
WOOD	.000	
FIBER	-1.803	.000

TWO-SIDED PROBABILITIES USING NORMAL APPROXIMATION

	WOOD	**FIBER**
WOOD	1.000	
FIBER	.071	1.000

Table 11.4
Wilcoxon results

The boat with the fiberglass daggerboard beat the matched boat with the wooden daggerboard twelve times, while the boat with the wooden daggerboard beat its matched pair eight times. Note that with this data the two-tailed probabilities are .071. This is all the information you need to make your decision. For this one-tailed example, you divide the .071 in half giving a p value of .0355.

5. Decide whether or not to reject the null hypothesis.

Because the p value of .0355 is less than the alpha value of .05, reject the null hypothesis.

6. Write a summary statement.

Because MYSTAT doesn't provide the Wilcoxon value, it cannot be included in the summary statement. However, MYSTAT does calculate a z value for this test. In this case, the following summary statement might be written: "The new fiberglass daggerboard significantly improved the performance of the racers when compared to the older wooden models ($z = -1.803$, $p = .0355$)."

Friedman Two-way ANOVA

The Friedman analysis of variance test is the nonparametric alternative to a repeated-measures analysis of variance. The Friedman test statistic is used when you wish to compare more than two dependent samples or measures. It is often used to analyze the rankings from multiple judges.

The FRIEDMAN command computes a Friedman nonparametric analysis of variance on selected variables. The Friedman test in MYSTAT averages tied ranks. The general form of the command is

FRIEDMAN [<VAR1>, <VAR2>, <...>]

As with the other nonparametric commands, if no variable names are typed after the Friedman command, the statistical test is conducted on all quantitative variables.

The Problem

You are the director of research for a large marketing company. Your corporation has produced four trial versions of a new nonalcoholic beer commercial. The commercials are targeted at college-age people. You randomly select twenty college students and ask them to rank the four commercials. The dependent variable is of rank order, and because all four measures come from the same person, it is a repeated measures problem. Because there are four different measures, the Friedman two-way ANOVA is used instead of the Wilcoxon procedure.

The Solution

1. Write the null and alternative hypotheses.

H_0: There are no systematic differences in the rankings across the judges.

H_1: There are systematic differences in the ranking across the judges.

2. Set the alpha level.

We'll use .05 again.

3. Collect the data.

The data are provided in the file NONALCOH.

4. Calculate the statistic.

1. Type **USE B:NONALCOH** and press [**Enter**] twice.
2. Type **FRIEDMAN** and press [**Enter**].

   The results are shown in Table 11.5.

FRIEDMAN TWO-WAY ANALYSIS OF VARIANCE RESULTS FOR 20 CASES

VARIABLE	RANK SUM
AD1	29.000
AD2	48.000
AD3	46.000
AD4	77.000

FRIEDMAN TEST STATISTIC = 35.7000
KENDALL COEFFICIENT OF CONCORDANCE = .595
PROBABILITY IS .000 ASSUMING CHI-SQUARE DISTRIBUTION WITH   3 DF

Table 11.5
Friedman two-way
ANOVA results

First, each variable is listed with the sum of its ranks. Remember that each of the twenty judges ranked these four commercials. If there had been one commercial that every judge picked to be best, the sum would be 20. On the other hand, if there had been one commercial that every judge ranked last, the sum would be 80. Looking at the rank sum, note that commercial one (AD1) is close to the lowest possible score, and AD4 is close to the highest possible score. Next comes the Friedman test statistic with a value of 35.7. Following the Friedman is the Kendall coefficient of concordance, which is an estimate of the average correlation among the judges' ratings of the commercials. Finally, the probability that you use to make your decision is reported.

5. Decide whether or not to reject the null hypothesis.

With a reported probability of .000 (which is less than .0005), you reject the null hypothesis.

6. Write a summary statement.

You might write the following summary statement: "The twenty judges rank-ordered the four advertisements differently (Friedman test statistic = 35.7, p < 0.0005). Advertisement 1 had the best average sum of ranks of 29. Its average ranking by the twenty subjects in this study was 1.45." (The average ranking is calculated by dividing the rank sum of 29 by the number of judges.)

————— Exercises

1. A consumer-research organization asked twenty-five men on their college's intramural soccer team to evaluate the effectiveness of three soccer training films. After viewing each film, each man indicated the film he preferred. Using the results found in the file SOCCER, determine whether there is a significant preference for any of the three films.

2. During the 1960s, a major university wanted to know whether there was a relationship between students' CLASS standing (freshmen = 1, sophomore = 2, junior = 3, senior = 4) and their political AFFILiation (Democrat = 1, Republican = 2). One hundred students were surveyed and the results are tabulated in the file UNIVER. Are class standing and political affiliation independent of each other using an alpha level equal to .01? Write a summary statement for this experiment.

3. Another experiment took place during the 1960s to see whether a series of lectures on social awareness could influence white students in dormitories to change their attitudes about rooming with minority students. Twenty students were matched on several environmental variables that were presumed to be related to the students' prior exposure to minority cultures. One member of each pair was randomly assigned to the social awareness lectures and the other member participated in a control activity. At the end of the experiment their answers to whether they were willing to live with a minority student next year were collected. Only the text output is contained in the file SOCIAL. You will need to use the MYSTAT Editor to change this text information into a form that can be analyzed by MYSTAT. Use the Sign test and an alpha level of .05 to evaluate these results. Did the lecture series make a difference? Write a summary statement.

4. A researcher wanted to know whether attractiveness was related to the decision to attend college after high school. The researcher collected

attractiveness scores on a scale of 1 through 10 for twelve sets of male cousins, one of whom went to COLLEGE and the other who went to WORK. Use the Wilcoxon test to determine whether there is a significant difference between these cousins. Write a summary statement. The data are in the file ATTRACT.

5. Fifteen subjects were exposed to four different learning environments. The first environment gave positive rewards for correct responses. The second gave punishments for incorrect responses. The third environment terminated a loud noise when a correct response was given (negative reinforcement). The fourth environment was a control and no action was taken when the subject responded. Subjects were given the different treatments in different orders and the amount of information retained after the end of the experiment was measured. The measurement scale used is of rank order. Use the Friedman analysis of variance by ranks to evaluate the data contained in the file LEARNING. Note that the variables (ONE, TWO, THREE, FOUR) refer to the score in each of the four environments. Were there differences in the amount of information retained across the different treatments? Write a summary statement for this experiment.

Appendix

Troubleshooting Guide

The Troubleshooting Guide is a list of error messages that you may receive when using MYSTAT. Most of these messages are produced by MYSTAT, but some are produced by DOS. The error messages are listed alphabetically (messages that start with *ERROR* are listed according to the first letter of the first word after ERROR). The possible cause or causes for the error message are given along with solutions.

In addition to the error messages listed here, there are other error messages that MYSTAT reports. These are specific in pointing out a probable cause for the error; for example, "You cannot DELETE before cases have been entered." Many of these error messages recommend courses of action that should fix the problem.

Occasionally, you may find that MYSTAT's assessment of the problem and the recommended solution don't help you. In these cases, try restarting the program and carefully repeating the operation you were attempting when you received the message.

Message	Cause	Solution
No message—program simply freezes.	This may occur when you are naming variables in the MYSTAT Editor. There is not enough available computer memory for MYSTAT to function properly.	Quit and restart the MYSTAT program, then reenter the MYSTAT Editor.
Cannot load overlay: too many open files. (User is returned to the DOS prompt.)	Your system does not have a CONFIG.SYS file, or your CONFIG.SYS file contains a missing or incorrect "FILES=..." line.	Make sure you have a CONFIG.SYS file and that it contains a FILES setting greater than 20. (See your lab instructor if you need help with this.) Restart MYSTAT after correcting the problem.
Error on writing to unit X— probably due to insufficient disk. (X is some number, usually 4. User is returned to the DOS prompt.)	The disk from which you are running MYSTAT does not have enough space to hold the temporary file that MYSTAT creates when it uses a data file.	Exit to DOS and restart MYSTAT. Make sure you are following the instructions in Chapter 1 for starting MYSTAT on your system (e.g., if you are running MYSTAT on a dual-drive system, you must start MYSTAT from the drive containing your data disk, usually drive B.)
Incorrect file assignment. No assignment made.	Not typing a filename after a drive specification.	Retype the command and include the filename after the drive specification.
No paper error writing device PRN Abort, Retry, Ignore, Fail? **Not ready error writing device PRN Abort, Retry, Ignore, Fail?**	Entering OUTPUT @ without a printer connected and ready, then tried to conduct a statistical test or calculation.	Type **A** to select abort. (If your system does not respond, see your lab instructor for help.) Connect a printer to your computer and make sure it is ready, then restart MYSTAT.
Unable to tell what you mean somewhere about here:	This is a message that MYSTAT is likely to produce when you enter a command that contains a typing error. MYSTAT places an up carat (^) directly below the point at which it could not understand your input. MYSTAT cannot process the command, not even to the extent that it can tell you what's wrong.	Look for omitted or mistyped characters in the command you entered (e.g., a missing quote before text values). Then look up the syntax and function of the command (or series of commands) and make sure you understand what it does and how.
Variable name already exists in file.	Trying to name a variable with a name that already exists.	Rename the variable with a unique name.

Message	Cause	Solution
Write fault error writing device PRN. Abort, Retry, Ignore, Fail?	Entering OUTPUT @ without a printer connected and ready, then tried to conduct a statistical test or calculation.	Type A to select abort. (If your system does not respond, see your lab instructor for help.) Connect and a printer to your computer and make sure it is ready, then restart MYSTAT.
You are not ready to read data. Be sure you have correctly entered USE sentences.	Trying to manipulate a file before entering USE X:<filename> where X is the drive reading the data files.	Type USE X:<filename> where X is the drive reading the data files.
You are trying to process the wrong kind of data for this procedure.	Trying to do a statistical test on character data.	Make sure the variables on which you are conducting the test are numeric.
You are trying to read an empty file.	(1) Omitting a drive specification in a command that requires one (e.g., USE <filename> or EDIT <filename>) or mistyping the filename. (2) Trying to run the Demo program on a dual-drive system when you started MYSTAT from the drive containing the data disk.	Retype the command and include the correct drive and filename. (e.g., USE B:<filename>) Refer to the section "The DEMO Command" on page 12 in Chapter 1.
You can enter commands only on line below window.	Entering a command while still in the Edit window. You must be on the command line to enter a command.	Press [Esc] to move the cursor to the command line, then reenter the command.
You can move cursor and enter data only from inside window.	Entering or editing data while still on the Command line. You must be in the Edit window to enter or edit data.	Press [Esc] to move the cursor to the Edit window, then reenter the data.
You entered data which do not match the variable's data type.	Trying to type a text value in a column designated for numeric data, or you tried to enter a numeric value in a column designated for character data.	Type a data value of the correct type.
You have not given an input file with USE command.	Trying to manipulate a file before entering USE X:<filename> where X is the drive reading the data files.	Type USE X:<filename> where X is the drive reading the data files.
You have used up your EDITOR variables space.	There is not enough available computer memory for MYSTAT to function properly.	Quit and restart the MYSTAT program, then reenter the MYSTAT Editor.
Write protect error reading drive X Abort, Retry, Fail?	Attempting to save or manipulate a file on a diskette which is write-protected.	Make sure the disk is not write-protected, then press **R** to retry.

References

Cleveland, W.S. 1985. *The elements of graphing data*. Monterey, CA: Wadsworth Advanced Books.

Huff, D. 1954. *How to lie with statistics*. New York: Norton.

Judd, C.M., and McClelland, G.H. 1989. *Data analysis: a model-comparison approach*. San Diego: Harcourt Brace Jovanovich.

Marascuilo, L.A., and McSweeney, M. 1977. *Nonparametric and distribution-free methods for the social sciences*. Belmont, CA: Wadsworth Publishing.

Neter, J., Wasserman, W., and Kutner, M.H. 1990. *Applied linear statistical models*. 3d ed. Homewook, IL: Irwin.

Pedhazur, E.J. 1982. *Multiple regression in behavioral research*. 2d ed. New York: Holt, Rinehart and Winston.

Siegel, S. 1956. *Nonparametric statistics for the behavioral sciences*. New York: McGraw-Hill.

Index

Symbols

A

B

C

Using Software

A Guide to the Ethical and Legal Use of Software for Members of the Academic Community

Software enables us to accomplish many different tasks with computers. Unfortunately, in order to get their work done quickly and conveniently, some people justify making and using unauthorized copies of software. They may not understand the implications of their actions or the restrictions of the U.S. copyright law.

Here Are Some Relevant Facts:

1. Unauthorized copying of software is illegal. Copyright law protects software authors and publishers, just as patent law protects inventors.

2. Unauthorized copying of software by individuals can harm the entire academic community. If unauthorized copying proliferates on a campus, the institution may incur a legal liability. Also, the institution may find it more difficult to negotiate agreements that would make software more widely and less expensively available to members of the academic community.

3. Unauthorized copying of software can deprive developers of a fair return for their work, increase prices, reduce the level of future support and enhancement, and inhibit the development of new software products.

Respect for the intellectual work and property of others has traditionally been essential to the mission of colleges and universities. As members of the academic community, we value the free exchange of ideas. Just as we do not tolerate plagiarism, we do not condone the unauthorized copying of software, including programs, applications, data bases, and code.

Therefore, we offer a statement of principle about intellectual property and the legal and ethical use of software. This "code"—intended for adaptation and use by individual colleges and universities—was developed by EDUCOM. The EDUCOM code is on the last page of this book.

THE EDUCOM CODE: Software and Intellectual Rights

Respect for intellectual labor and creativity is vital to academic discourse and enterprise. This principle applies to works of all authors and publishers in all media. It encompasses respect for the right to acknowledgment, right to privacy, and right to determine the form, manner, and terms of publication and distribution.

Because electronic information is volatile and easily reproduced, respect for the work and personal expression of others is especially critical in computer environments. Violations of authorial integrity, including plagiarism, invasion of privacy, unauthorized access, and trade secret and copyright violations, may be grounds for sanctions against members of the academic community.

Questions You May Have About Using Software

a. What do I need to know about software and the U.S. Copyright Act?

Unless it has been placed in the public domain, software is protected by copyright law. The owner of a copyright holds exclusive right to the reproduction and distribution of his or her work. Therefore, it is illegal to duplicate or distribute software or its documentation without the permission of the copyright owner. If you have purchased your copy, however, you may make a backup for your own use in case the original is destroyed or fails to work.

b. Can I loan software I have purchased myself?

If your software came with a clearly visible license agreement, or if you signed a registration care, READ THE LICENSE CAREFULLY before you use the software. Some licenses may restrict use to a specific computer. Copyright law does not permit you to run your software on two or more computers simultaneously unless the license agreement specifically allows it. It may, however, be legal to loan your software to a friend temporarily as long as you do not keep a copy.

c. If software is not copy-protected, do I have the right to copy it?

Lack of copy-protection does NOT constitute permission to copy software in order to share or sell it. "Non-copy-protected" software enables you to protect your investment by making a backup copy. In offering non-copy-protected software to you, the developer or publisher has demonstrated significant trust in your integrity.

d. May I copy software that is available through facilities on my campus so that I can use it more conveniently in my own room?

Software acquired by colleges and universities is usually licensed. The licenses restrict how and where the software may be legally used by members of the community. This applies to software installed on hard disks in microcomputer clusters, software distributed on disks by a campus lending library, and software available on a campus mainframe or network. Some institutional licenses permit copying of certain purposes. Consult your campus authorities if you are unsure about the use of a particular software product.

e. Isn't it legally "fair use" to copy software if the purpose in sharing it is purely educational?

No. It is illegal for a faculty member or student to copy software for distribution among the members of a class, without permission of the author or publisher.

Restrictions on the use of software are far from uniform. You should check carefully each piece of software and the accompanying documentation yourself. In general, you do not have the right to:

1. receive and use unauthorized copies of software, or
2. make unauthorized copies of software for others.

This information is from a copyrighted brochure by EDUCOM, a non-profit consortium of over 600 colleges and universities committed to the use and management of information technology in higher education, and ADAPSO, the computer software and services industry association.

EDUCOM
EUIT
1112 16th St., NW
Suite 600
Washington, DC 20036

ADAPSO
1616 N. Fort Myer Drive
Suite 1300
Arlington, VA 22209-9998